MAIZE TECHNOLOGY DEVELOPMENT AND TRANSFER

A GIS Application for Research Planning in Kenya

To Mekki and Dihaiba

Maize Technology Development and Transfer

A GIS Application for Research Planning in Kenya

Edited by

R.M. Hassan

Department of Agricultural Economics
University of Pretoria
South Africa

CAB INTERNATIONAL
in association with the
International Maize and Wheat Improvement Center (CIMMYT)
and the
Kenya Agricultural Research Institute (KARI)

CAB INTERNATIONAL
Wallingford
Oxon OX10 8DE
UK

Tel: +44 (0)1491 832111
Fax: +44 (0)1491 833508
Email: cabi@cabi.org

CAB INTERNATIONAL
198 Madison Avenue
New York, NY 10016
USA

Tel: +1 212 726 6490
Fax: +1 212 686 7993
Email: cabi-nao@cabi.org

A catalogue record for this book is available from the British Library, London, UK

Library of Congress Cataloging-in-Publication Data

Maize technology development and transfer : a GIS application for research planning in Kenya / edited by R.M. Hassan.
p. cm.
Includes index.
ISBN 0-85199-287-0 (alk. paper)
1. Corn--Technological innovations--Kenya. 2. Agriculture--Technology transfer--Kenya. 3. Geographic information systems--Kenya. 4. Corn--Technological innovations--Research--Kenya. 5. Agriculture--Technology transfer--Research--Kenya. 6. Geographic information systems--Research-- Kenya. I. Hassan, Rashid M. II. International Maize and Wheat Improvement Center. III. Kenya Agricultural Research Institute.
SB191.M2M325 1998
633.1′5′096762–dc21

98-14961
CIP

ISBN 0 85199 287 0

Published in association with

International Maize and Wheat Improvement Center (CIMMYT)
Apdo. Postal 6-641
Mexico DF
Mexico

and

Kenya Agricultural Research Institute (KARI)
PO Box 57811
Nairobi
Kenya

Typeset in 10/12 Garamond by York House Typographic Ltd, London
Printed and bound in the UK at the University Press, Cambridge

Contents

Contributors

John D. Corbett is a Geographer and Research Scientist with the Integrated Information Management Laboratory, Blackland Research Center, Texas A&M University System, Temple, Texas, USA.

Rashid M. Hassan is a Professor with the Department of Agricultural Economics, University of Pretoria, South Africa.

Geofry Kamau is a Maize Agronomist with the Regional Research Center at Mtwapa, Kenya Agricultural Research Institute.

Danial D. Karanja is an Agricultural Economist with the National Plant Breeding Centre at Njoro, Kenya Agricultural Research Institute.

A. Laboso is a Maize Breeder with the National Agricultural Research Centre at Kitale, Kenya Agricultural Research Institute.

John Lynam is a Senior Scientist, Agricultural Sciences Division, with the Rockefeller Foundation, Nairobi, Kenya.

Bradford Mills is an Assistant Professor, Department of Agricultural Economics, Virginia Polytechnic Institute and State University, Blacksburg, Virginia, USA.

H.H.A. Mulamula is the Director of the Regional Research Centre at Kakamega, Kenya Agricultural Research Institute.

Festus Muriithi is an Agricultural Economist with the Regional Research Centre at Embu, Kenya Agricultural Research Institute.

Peterson Mwangi is an Agricultural Economist with the National Agricultural Research Laboratories, Kenya Agricultural Research Institute, Nairobi.

Wilfred Mwangi is Principal Economist for Eastern Africa with the International Maize and Wheat Improvement Center, Addis Ababa, Ethiopia.

Mugo Njore is a Maize Breeder with the National Dry Land Farming Research Center at Katumani, Kenya Agricultural Research Institute.

Kiarie Njoroge is a Maize Breeder and National Coordinator of the Maize Research Programme, Kenya Agricultural Research Institute, Nairobi.

Peter Okoth is a Soil Scientist with the Kenya Soil Survey, Kenya Agricultural Research Institute, Nairobi.

Ruth Onyango is a Maize Agronomist with the National Agricultural Research Centre at Kitale, Kenya Agricultural Research Institute.

Robin Otsyula is a Maize Breeder with the Regional Research Center at Kakamega, Kenya Agricultural Research Institute.

Joel K. Ransom is a Maize Agronomist with the International Maize and Wheat Improvement Center, Nairobi, Kenya.

J.K. Rutto is the Deputy Director of the Kenya Agricultural Research Institute, Nairobi, Kenya.

Foreword

More than three decades of systematic research have been directed toward maize technology development and transfer in Kenya. During that period, the maize research program of the Kenya Agricultural Research Institute (KARI) developed successful technologies that profoundly affected maize production, both within Kenya and elsewhere in eastern Africa. The rapid diffusion of improved maize technologies in Kenya is a well-known African parallel to the spread of hybrid maize in the USA. This is especially true in the highlands of Kenya, where most maize land was sown to hybrids within less than two decades from the birth of formal maize research at Kitale.

However, the past three decades have also witnessed significant change in the socioeconomic and institutional environment of farming in Kenya, presenting newer challenges to maize research, extension, and policy. Land reform, population growth and movement, economic and infrastructural development – among other trends – have shaped corresponding adaptive responses in maize production. The maize research program has continued to adjust its emphasis and efforts accordingly. Intent on obtaining better guidance for setting its research priorities, KARI was particularly interested in exploiting the new tools of research evaluation, which would make it possible to combine and analyze many different kinds of data related to Kenya's maize economy. As a result, the Kenya Maize Data Base Project (MDBP) was launched in collaboration with the Economics Program of the International Maize and Wheat Improvement Center (CIMMYT). The MDBP, which formed part of the national program of maize research in Kenya, received partial funding from the Rockefeller Foundation and the bilateral USAID mission in Kenya for two years, starting March 1992.

Before the MDBP was established, KARI had not engaged in such a comprehensive, country-wide effort to assess its record of success and determine the future directions for maize research. Very rapidly, during the project's first year, the significance of its potential contribution to methods of effective research planning and evaluation became evident. The project was thus extended into a third year through a Rockefeller Foundation grant.

This volume contains detailed discussions of the approach and methods used, and the results and lessons learned, from the Kenya MDBP. With the exception of a very few recent attempts – mainly at international agricultural research institutions – the work documented in this volume represents a pilot endeavor to use advanced computer technologies and spatial analysis tools in a unique interdisciplinary manner in a national agricultural research center. One of the project's important innovative features was the development of procedures for integrating information from farmer interviews with data from biological experiments, secondary regional and national statistics on major socioeconomic phenomena, and other spatial and agroclimatic attributes into a single, spatially linked digital database. This integration was achieved by spatial analyses and geo-referencing of data. To target prospective technological innovations accurately to farmers' conditions, an improved definition of maize production environments was needed. Geographic information system (GIS) techniques were used to characterize agroclimatic diversity and delineate maize-specific adaptation zones. This effort provided the base map onto which other layers of information were superimposed – for example, information on population density. The union of this information produced a powerful, dynamic database and planning tool, which enabled KARI to establish, for the first time with high accuracy, the extent of maize cultivation in each zone. Important correlations were eventually revealed between the intensity of maize cultivation, distribution of human population, climate, and economic infrastructure.

These patterns of spatial correlation guided the design of a survey intended to integrate data on farmers' attributes into the database. An innovative sampling frame, which combined elements of both spatial and population-based sampling, was developed for conducting extensive country-wide surveys. Neither KARI nor any other research or statistics agency in Kenya or the region has such a sampling frame for maize. The resulting survey data were paired with agroclimatic and other spatial data to characterize maize cropping systems and refine the definition of adaptation zones, which led to the identification of important deficiencies in the old zonation scheme. A detailed account of this process is given in the methods chapters (Chapters 2, 3, and 4), which constitute Part I of the book.

Part II of this volume presents the results of different approaches to *ex ante* research evaluation and setting of research priorities for maize using the database and spatial framework. In Chapter 5, data on farmers' perceptions of important constraints to increased productivity were combined with spatial attributes (climate, area, population) to rank zones, constraints, and research themes using economic efficiency and nonefficiency criteria. These rankings were compared with the historical allocation of maize research resources to identify departures from farmers' assessments. Chapter 6 takes this analysis further, extending the agroecological approach to an application of economic surplus methods for setting research priorities across multiple, spatially linked maize production zones.

Part III presents the various ways in which the integrated database was used in analyzing patterns of maize technology diffusion and impact and establishing a reference point for monitoring the process of technological transformation in maize production in Kenya. Determinants of the spatial diversity and temporal variations in adoption of improved maize seed and fertilizer and the resulting productivity differentials and unexploited yield gaps are examined in Chapters 7 and 8. In Chapter 9, the spatial framework and database are used to examine where maize technology could make the greatest impact in countering infestations of the parasitic weed, *Striga*, in maize. Finally, the role of extension and other agricultural support services in promoting modern maize technologies is analyzed in Chapter 10.

The final part of this book synthesizes the major findings and implications of the MDBP research. We believe that the results of this work have significantly altered the way maize research is organized in Kenya. Methods developed through the MDBP have already been extended to other commodity and factor research programs within KARI. There is also growing interest in replicating this approach for the planning and evaluation of agricultural research in other countries in eastern and southern Africa. We fully expect that the methods and lessons of the MDBP will continue to be a rich source of information on a wide range of applications for biological as well as socioeconomics research themes and questions.

Cyrus Ndiritu

Director, Kenya Agricultural Research Institute

Acknowledgments

The work presented in this volume is a product of the combined efforts, knowledge, and dedication of many individuals and institutions. During the early 1990s, the Economics Program of the International Maize and Wheat Improvement Center (CIMMYT) gave priority to developing methods for forward-looking evaluations of agricultural research and technology design. This marked an important shift of focus to priority setting and *ex ante* research evaluation and planning in the Program's agenda. At the same time, the Kenya Agricultural Research Institute (KARI) was addressing the challenge of reviving Kenya's early success in transforming the country's maize subsector. The impressive productivity gains achieved over the past two decades in maize, the country's basic food staple, were no longer sufficient to meet the demand for maize from a rapidly growing population. Accordingly, KARI and CIMMYT agreed to apply emerging tools of research evaluation and spatial technology to the case of maize technology research in Kenya. The proposal found strong interest from the Rockefeller Foundation, which supported similar country-specific and Africa-wide initiatives. This confluence of interests led to the birth of the Kenya Maize Data Base Project (MDBP) and of this book, under partial funding from the Rockefeller Foundation and the Kenya Mission of the US Agency for International Development (USAID). I am very grateful to both institutions for the financial assistance. Special thanks go to Dr G. Gingerich of USAID, who initiated funding arrangements together with Dr J. Lynam of the Rockefeller Foundation. The support provided to this project by Dr Lynam extended far beyond provision of financial assistance. Every member of this project is indebted to him for invaluable, direct technical and professional input throughout the two and a half years that we worked together. Dr Lynam was instrumental in getting support for the publication of this work, which will enable many people who are unfamiliar with the project to understand its objectives, achievements, and potential throughout Africa.

I would also like to express my gratitude to Dr C. Ndiritu, the Director of

KARI, and his staff, whose constant support of the Maize Data Base Project was crucial to its success. The significant contributions of Drs D. Byerlee and R. Tripp of the CIMMYT Economics Program to the design and implementation of this project in its early stages are much appreciated. Support from CIMMYT continued under Dr L. Harrington, who gave special attention to the completion of this work. I am also indebted to Dr J. Corbett, P. Okoth, and P. Maingi for their help with the geographic information system (GIS) work and design of the spatial sampling frame. The MDBP was implemented as an integral component of the maize and socioeconomics research programs at KARI. Without the great contributions of the large number of maize researchers and socioeconomists involved, completion of this work would have been impossible. I would like to thank them all, especially Dr K. Njoroge, the National Coordinator of KARI's maize research program and Chairman of the KMDBP National Steering Committee, and Dr D. Karanja, the project counterpart, for their critical assistance and support in the course of the project. The assistance and technical input of Dr A. Low, who participated in some of the planning workshops at KARI's regional research centers, is highly appreciated as well. Finally, the patience and generous help provided by the maize farmers, from whom I learned a great deal about maize production in Kenya, helped lay the foundations for the information and knowledge summarized in this volume. Without them, this book could not have been written.

Many other individuals contributed in various ways to this work, from participation in planning and technical consultative workshops to reviewing manuscripts of the chapters of this book. Those whose names have not been mentioned already include G. Edmeades, P. Ewel, S. Franzel, P. Heisey, D. Jewell, A. Mbabu, B. Mills, L. Muhammad, W. Mwangi, J. Qureshi, J. Ransom, M. Smale, S. Waddington, and B. Wafula. I cannot thank Kelly Cassaday enough for the patient editorial assistance she provided in the long process of producing this volume.

Last but never least, I would like to thank my wife, Entisar, and sons Mekki, Hatim, and Mohammed for their love, motivation, and the support they have generously provided, suffering my long absence from home on this work and sacrificing dear times of family life.

Rashid M. Hassan

Acronyms and Abbreviations

BLUE	Best linear, unbiased estimators
CBS	Central Bureau of Statistics (Kenya)
CIMMYT	Centro Internacional de Mejoramiento de Maíz y Trigo (International Maize and Wheat Improvement Center)
COSCA	Collaborative Study of Cassava in Africa
DEM	Digital elevation model
DRSRS	Department of Remote Sensing and Resource Surveys (Kenya)
FURP	Fertilizer Use Recommendation Project
GCV	Generalized cross validation
GIS	Geographic information system
GPS	Global Positioning System
KARI	Kenya Agricultural Research Institute
MDBP	Maize Data Base Project
NARS	National agricultural research system
NASSEP	National Sample Surveys and Evaluation Program
NSF	National Sampling Frame
OLS	Ordinary Least Sqaures
OPV	Open-pollinated variety
SURE	Seemingly Unrelated Regressions Estimation

Geographical Data

Geographical Data Table 1. Initial classification of maize-specific zones: climatic attributes, size, and actual sampling fractions.

Zone	Elevation (masl)[a]	Total precipitation, March–August (mm)	Average temperature (°C), March–August[b] Min.	Max.	Mean	Maize area, 1990[c] (000 ha)	Percentage of total maize area	Percentage of farmers in sample
Lowland tropics	**< 1000**	–	**21.0**	**31.0**	24.6	**42**	**4.1**	**7.1**
Dry lowland tropics	–	**300–500**	20.4	29.0	24.0	13	1.3	2.8
Moist lowland tropics	–	**> 500**	21.1	29.4	25.4	29	2.8	4.3
Midaltitude zone	**1000–1800**	–	**16.0**	**28.0**	21.8	**566**	**54.6**	**53.8**
Dry midaltitude zone	–	**< 500**	14.2	25.8	22.7	110	10.6	10.0
Moist midaltitude zone	–	**> 500**	13.8	26.4	21.5	456	44.0	43.8
Highland tropics	**> 1800**	–	**8.0**	**24.0**	17.0	**408**	**39.3**	**33.5**
Dry highland tropics	–	**< 500**	10.6	24.2	16.9	100	9.6	7.5
Moist highland tropics	–	**> 500**	11.0	24.7	17.0	265	25.6	21.6
Cool highland tropics	**> 2000**	**< 1000**	**7.0**	**20.0**	15.0	**43**	**4.1**	**4.4**
Extreme water stress	**< 1000**	**< 300**	**19.0**	**30.0**	25.6	**21**	**2.0**	**5.6**
Total	–	–	–	–	–	**1037**	**100.0**	**100.0**

[a] masl = meters above sea level.
[b] Entries not in boldface type represent average values of attributes.
[c] Otichillo, W.K. and Sinange, R.K. (1991) *Long Rains Maize and Wheat Production in 1990.* Technical Report No. 140. Department of Resource Surveys and Remote Sensing, Ministry of Planning and National Development, Nairobi, Kenya.

Geographical Data Table 2. Revised classification of maize-specific zones: climatic attributes, size, and actual sampling fractions

Zone	Elevation (masl)[a]	Total precipitation, March–August[b] (mm)	Average temperature (°C), March–August Min.	Max.	Mean	Maize area, 1990[c] (000 ha)	Percentage of total maize area	Percentage of farmers in sample
Lowland tropics	–	–	–	–	–	**33**	**3.2**	**7.1**
Dry lowland tropics	**> 700**	**300–550**	**20.0**	**30.0**	25.4	12.5	1.2	2.8
Moist lowland tropics	**< 400**	**> 550**	**20.0**	**31.0**	25.8	21.5	2.0	4.3
Midaltitude zone	–	–	–	–	–	**197**	**19.0**	**20.2**
Dry midaltitude zone	**700–1400**	**300–550**	**14.0**	**33.0**	22.0	79	7.7	7.2
Moist midaltitude zone	**1110–1500**	**> 550**	**13.0**	**30.0**	22.1	118	11.3	13.0
Transitional zone	–	–	–	–	–	**461**	**44.5**	**35.0**
Dry transitional zone	**1100–1700**	**< 550**	**11.0**	**27.0**	19.7	37	3.6	5.7
Moist transitional zone	**1200–2000**	**> 500**	**11.0**	**29.0**	19.7	424	40.9	29.3
Highland tropics	–	–	–	–	–	**307**	**29.6**	**32.0**
Dry highland tropics	1600–2300	**< 550**	**8.0**	**26.0**	16.6	33	3.1	6.0
Moist highland tropics	1600–2700	**> 550**	**7.0**	**27.0**	16.7	238	23.0	21.6
Cool highland tropics	2000–2900	**< 1000**	**5.0**	**22.0**	13.8	36	3.5	4.4
Extreme water stress	**400–1100**	**< 400**	**16.0**	**32.0**	23.8	**39**	**3.7**	**5.7**
Total	–	–	–	–	–	**1037**	**100.0**	**100.0**

[a] masl = meters above sea level.
[b] Entries in boldface type represent criteria applied, whereas those not in bold represent average values of respective attributes in the zones.
[c] Otichillo, W.K. and Sinange, R.K. (1991) *Long Rains Maize and Wheat Production in 1990.* Technical Report No. 140. Department of Resource Surveys and Remote Sensing, Ministry of Planning and National Development, Nairobi, Kenya.

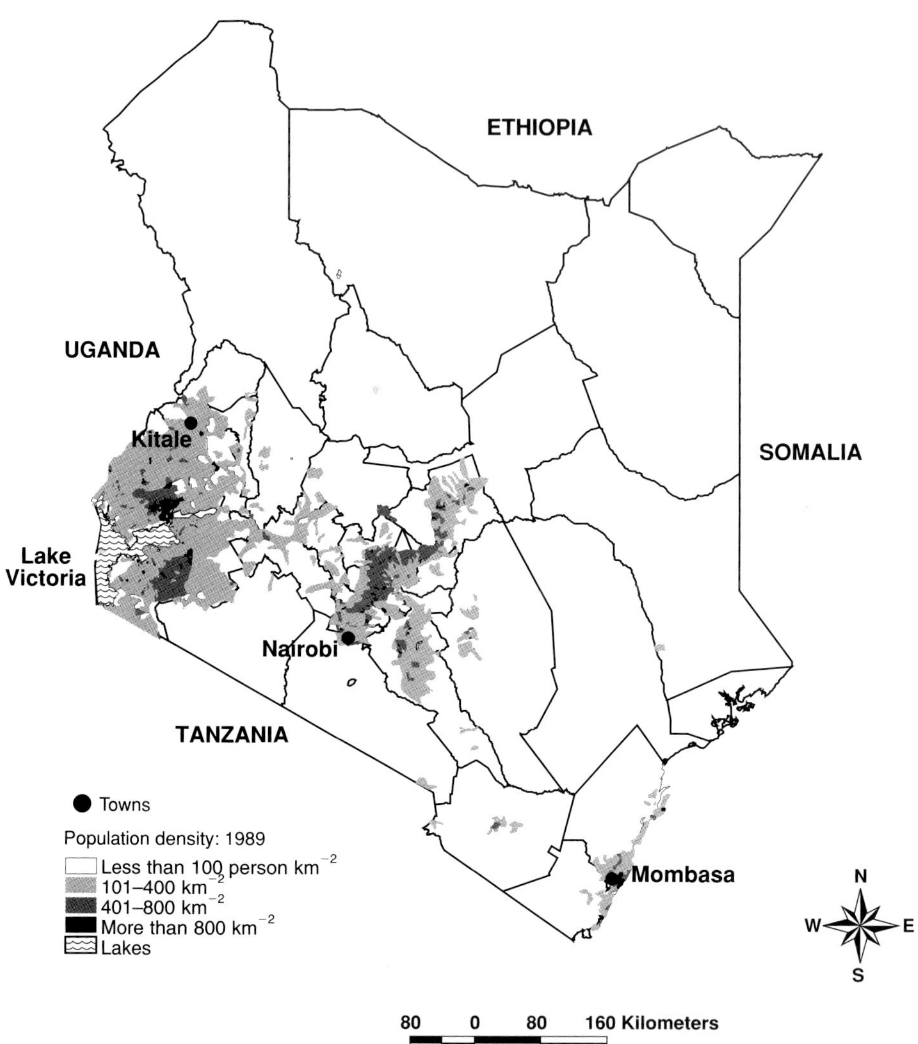

Map 1. Distribution and density of population in Kenya (1989 census).
Source: Kenya Maize Data Base Project (1992-93)

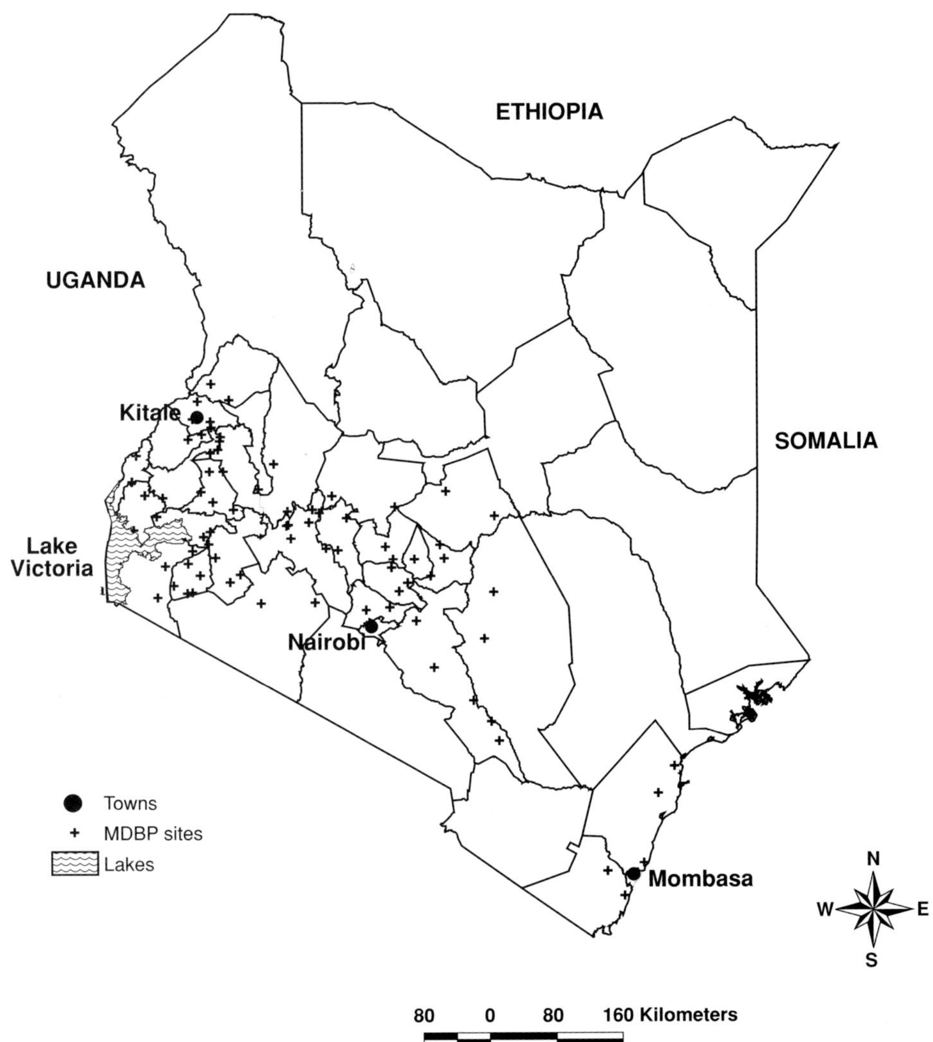

Map 2. Maize Data Base Project survey sites, 1992.
Source: Kenya Maize Data Base Project (1992-93)

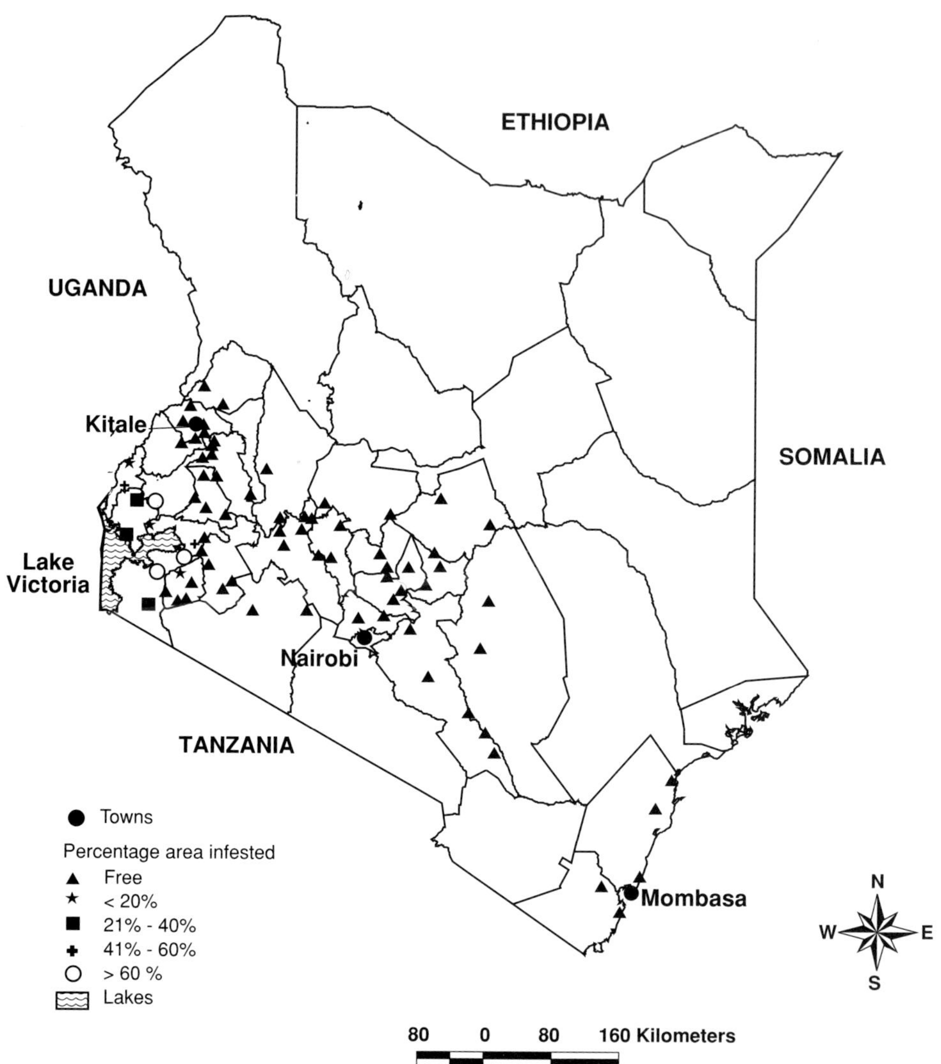

Map 3. Spatial distribution and severity of *Striga* infestation in maize in Kenya, 1992/93.
Source: Kenya Maize Data Base Project (1992-93)

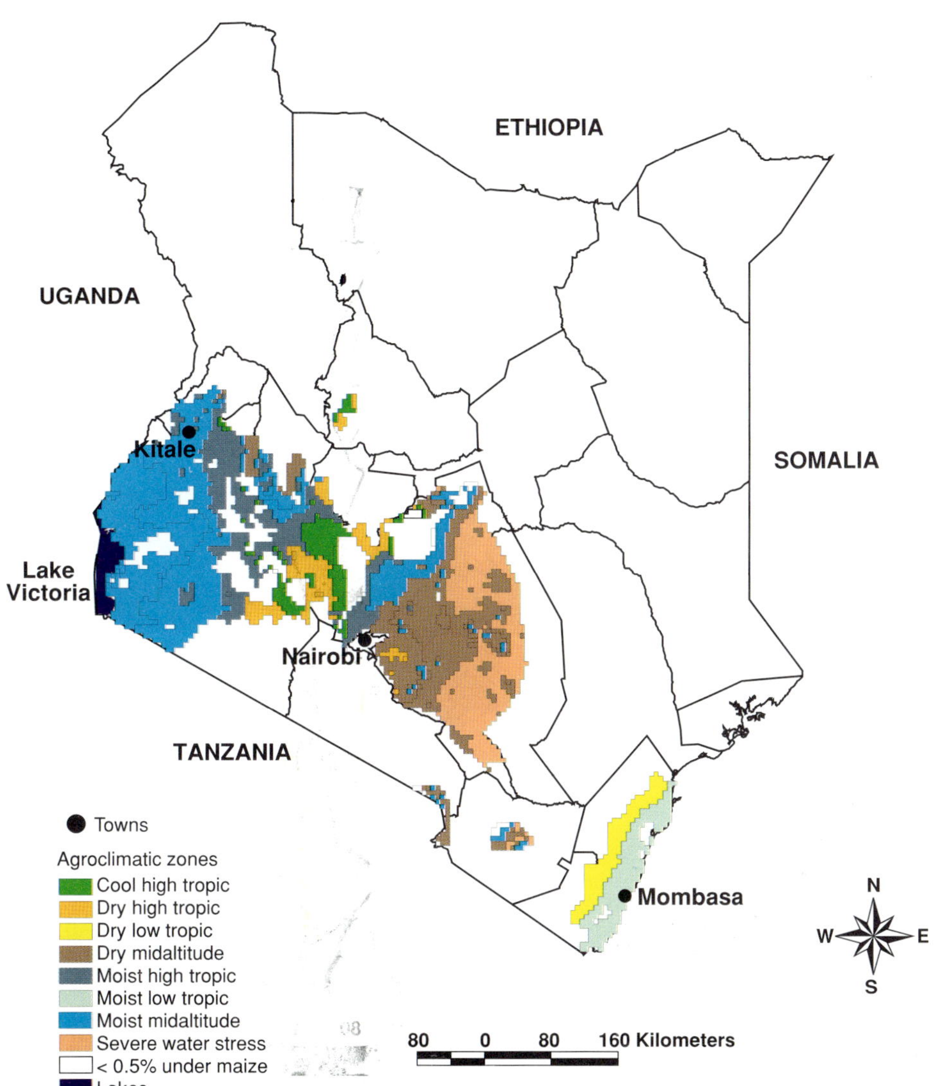

Plate 1. Initial zonation of maize production environments in Kenya.

Source: Kenya Maize Data Base Project (1992-93).

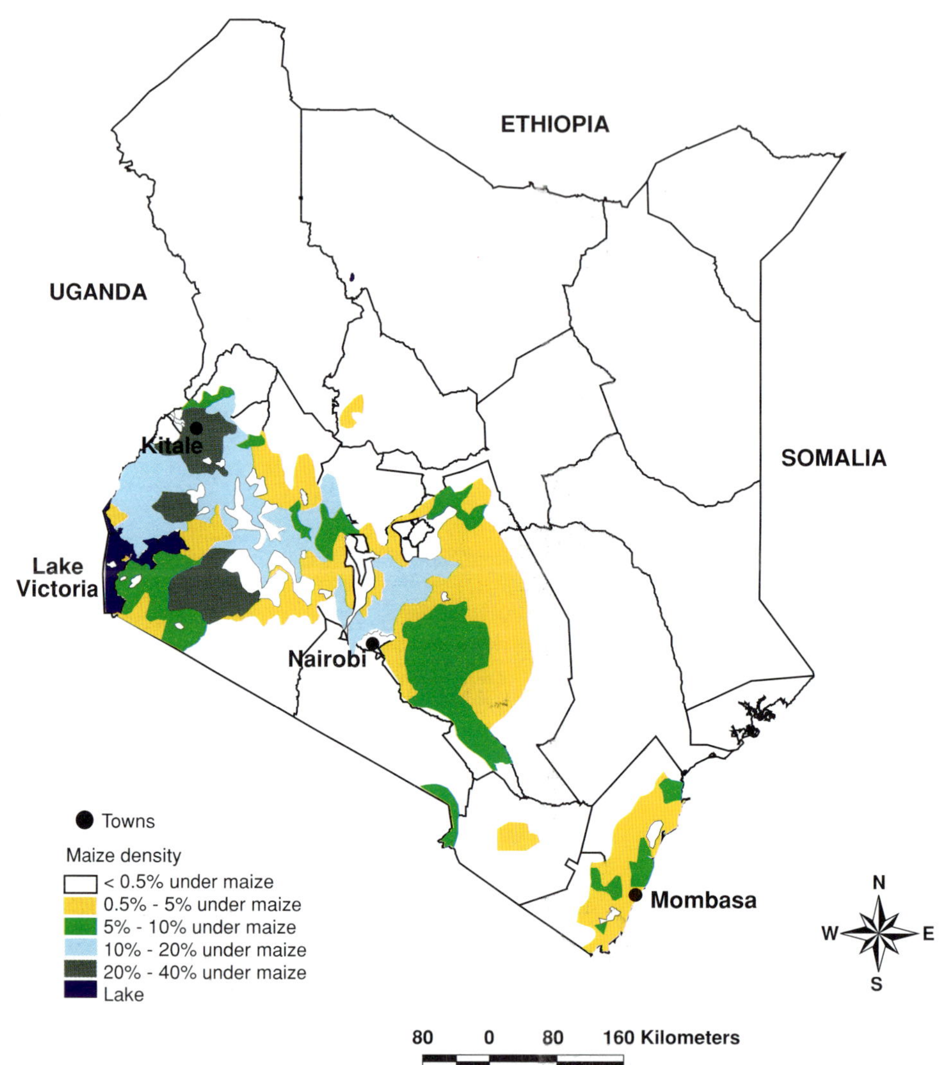

Plate 2. Intensity of maize cultivation in Kenya, 1989 long rains, DRSRS.

Source: Kenya Maize Data Base Project (1992-93).

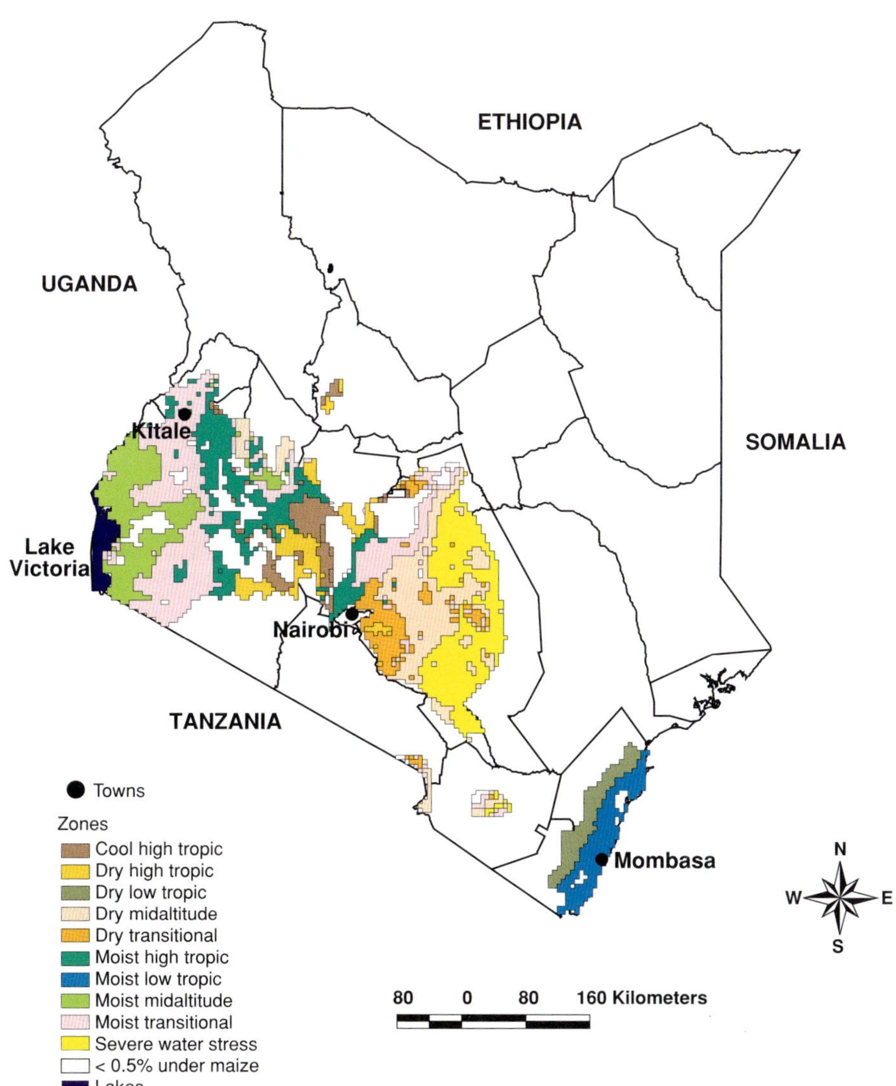

Plate 3. Revised intermediate agroclimatic zonation scheme for maize in Kenya.
Source: Kenya Maize Data Base Project (1992-93).

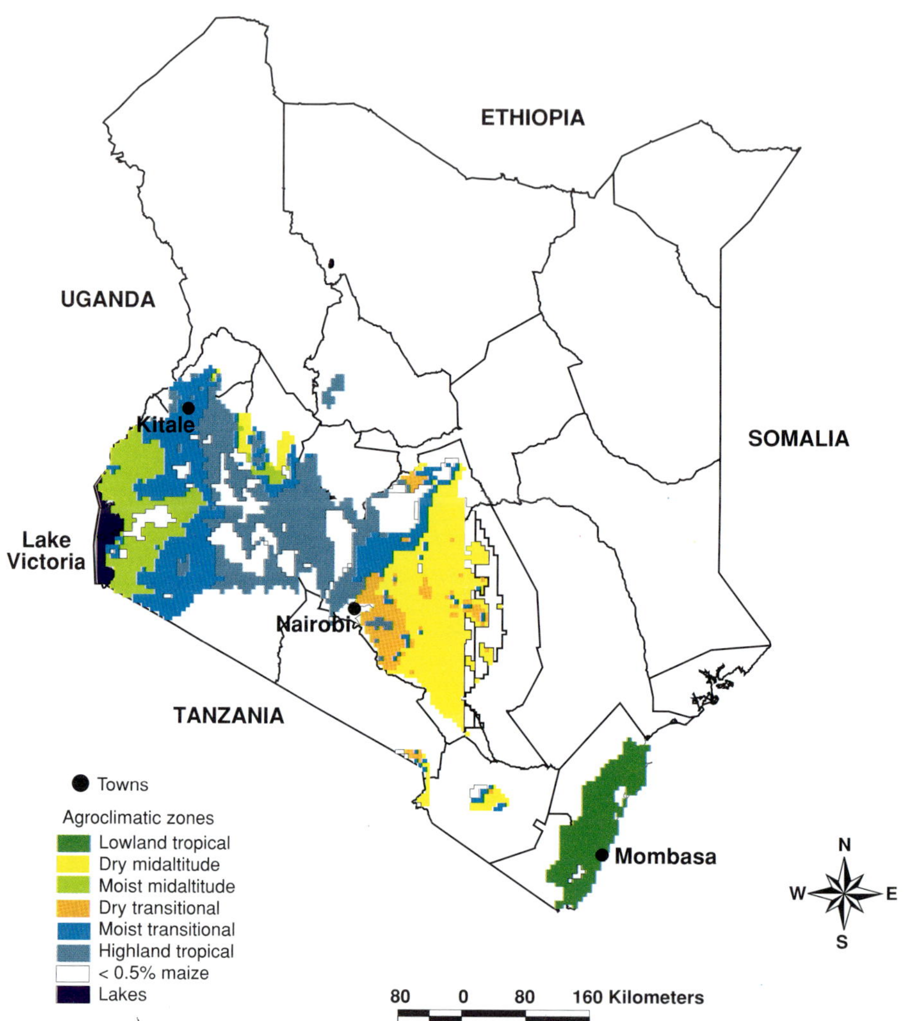

Plate 4. Final agroclimatic zonation of maize production environments in Kenya.
Source: Kenya Maize Data Base Project (1992-93)

I Introduction and Methods

1 A New Approach to Securing Sustained Growth in Kenya's Maize Sector

JOHN LYNAM AND RASHID M. HASSAN

To provide some background on the setting and objectives for the Kenya Maize Data Base Project, this chapter reviews the history and achievements of maize research in Kenya and describes the structure of the national maize research program. The strategic decisions that confront the national maize research program as it prepares to meet the needs of a future generation of Kenyan citizens are reviewed as well. Finally, the complex institutional issues surrounding the development and maintenance of an information system such as the Kenya maize database are described.

Introduction

The agricultural frontier is closing in Africa just as the continent enters its demographic transition, from which it will emerge with a population that is more urban than rural. The only way for Africa's burgeoning population to maintain per capita food supplies from a fixed land base will be to increase crop yields. With each passing year, agricultural research and improved technologies will become more necessary to assure Africa's ability to feed itself into the middle of the twenty-first century, yet so far the record of technology development and diffusion in Africa has been uneven at best. Research resources have declined so precipitously that public national agricultural research systems (NARSs) can scarcely maintain budgets and effective research programs (Pardey *et al.* 1995). For this reason, it is important to evaluate the research programs that have been successful in Africa and to analyze options for fostering continued growth in yields over the next 50 years. By identifying the conditions for successful research, NARSs across Africa will discover how to foster the effective research programs and appropriate technologies that are so desperately needed.

Fortunately, Kenya's experience of maize research and technology development opens a window onto the prerequisites for successful research in

Africa, revealing which factors promote sustained yield growth and which ones constrain growth. Through the Maize Data Base Project (MDBP), the Kenyan national research program has gained valuable insights into these issues in relation to maize research, expanded our understanding of their complexity, and developed new strategies for designing agricultural research in an African context. This book describes the methods and achievements of the MDBP in the belief that this project can serve as a model for other national research programs that seek to deploy their scarce resources more effectively.

However, before taking a closer look at the inner workings of the MDBP, it is important for readers to know something about the setting in which the project was implemented and the direction it seeks to define for future maize research. In this chapter, we review the history and achievements of maize research in Kenya and describe the structure of the national maize research program. Next, we discuss some of the strategic decisions that confront the national maize research program as it prepares to meet the needs of a future generation of Kenyan citizens. Finally, we examine the complex institutional issues surrounding the development and maintenance of an information system such as the Kenya maize database, so that it may continue contributing to enlightened decisions on research strategies.

Maize Research in Kenya: from Achievement to Adversity

Maize is not a traditional African crop. It arrived in East Africa with the Portuguese in the sixteenth century (Miracle 1966), but several centuries would pass before it became established as a major food staple in East Africa. The First World War created a considerable export market for maize, to which East Africa's white settlers and smallholders responded by expanding their production of maize. Internal demand for maize as a consumption item grew as well, especially when the traditional food crops, sorghum and millet, failed because of drought. By the end of the Second World War, the role of maize as a major food and cash crop in Kenya was consolidated. The government, recognizing the importance of the crop, initiated a maize improvement research program in 1955 at Kitale, the center of maize production in the White Highlands. Research at Kitale focused on developing late-maturing hybrids for the highland areas where rainfall was confined to one long season. Between 1957 and 1964, three additional maize research programs were established. The Katumani program concentrated on maize for semiarid mid-altitude areas; the Embu program, on moist midaltitude areas; and the Mtwapa program, on lowland, coastal areas. Rainfall is bimodal in all three environments.

By the mid-1970s, ten hybrid maize varieties and three composites had been released. Significant numbers of farmers adopted this first generation of improved maize materials (Gerhart 1975; Hesselmark 1976, unpublished paper on maize yields in Kenya, 1975, Maize and Produce Board, Nairobi). In many ways, the adoption of improved maize in Kenya mirrored the diffusion of improved rice and wheat varieties in Asia following the Green Revolution in the mid-1960s. The initial adopters tended to be large-scale farmers concen-

Table 1.1. Average growth rates in maize area, yield, and adoption of improved seed and fertilizer in Kenya, 1963–91.

	Growth rate			
	1963–74	1975–84	1985–91	Overall, 1963–91
Area (% yr^{-1})	4.6	0.4	(1.1)	2.3
Yield (% yr^{-1})	10.9	0.6	4.4	7.1
Number of new varieties released	13	2	6	21
Percentage of farmers who have adopted improved seed[a]				
Large-scale farmers in high-potential zones	47.5	24.2	22.2	93.9
Small-scale farmers in high-potential zones	16.0	42.2	36.6	94.7
Small-scale farmers in low-potential zones	4.0	12.1	39.7	56.8
Percentage of farmers who have adopted fertilizer[a]				
Large-scale farmers in high-potential zones	42.3	17.3	23.1	82.7
Small-scale farmers in high-potential zones	10.8	24.6	27.8	63.2
Small-scale farmers in low-potential zones	2.3	2.3	6.0	10.6

Source: Production, yield, and area data from the Ministry of Planning and National Development, *Statistical Abstracts.* Adoption data from the Kenya Maize Data Base Project farmer surveys (1992/93).

[a] Represents the percentage of farmers who once bought improved seed or fertilizer, but does not reflect current adoption rates. Large-scale farmers were defined as farmers having more than 8 ha of land. High-potential zones consist of the wet highlands and midaltitude regions. Low-potential zones are the semiarid and lowland tropical zones (Chapter 2).

trated in the high-potential areas; by 1974, about half of these farmers used the new maize materials. Small-scale farmers in the high-potential areas lagged behind the others in adopting the new varieties (16% had adopted them by 1974), but eventually they adopted them quite rapidly (58% by 1984) (Table 1.1). Small-scale farmers in the more marginal areas were the slowest to adopt improved maize, which was grown by only 16% of farmers in 1984. As in Asia, in Kenya the adoption of inorganic fertilizer followed closely on the adoption of improved seed in the large farm sector. But the Kenyan experience differed from the Asian experience in that smallholders' adoption of fertilizer lagged substantially behind their adoption of improved varieties and remained virtually negligible in marginal areas.

Despite these differences, this first generation of technical change in Kenyan maize production followed a Green Revolution path fairly closely. Farmers' yields increased largely because they used improved varieties, particularly hybrids, supplemented by purchased inputs, especially fertilizer. As in Zimbabwe, where improved maize materials had similarly been adopted fairly early (Eicher 1995), certain conditions were responsible for this success. First, research could considerably improve upon the maize varieties grown in Kenya, where new materials had been introduced in only limited numbers and only rather recently. Accordingly, a large yield gap was just waiting to be exploited through crop improvement (in fact, it has been difficult to surpass the original hybrid made by crossing Kitale Synthetic with Ecuador 573). Second, maize research in Kenya has received strong government and donor support since 1955, except for a brief lapse in the early 1980s. Third, formation of the Kenya Seed Company created a well-developed production and distribution system for hybrid seed. Fourth, Kenya's maize-producing areas are well served by a relatively dense road network, which permits effective marketing of grain and purchase of inputs. Finally, although maize pricing and marketing policy were controlled by state marketing boards until 1994,[1] the intent was to support maize production, at times through subsidies on seed and fertilizer. Since the Cereal Sector Reform Program began in 1988, the maize market has been liberalized gradually, although at times the disincentives for maize production have been great. In essence, these five conditions closely resemble the circumstances that sparked the success of modern rice and wheat varieties in Asia from the mid-1960s through the 1970s.

However, Kenya's maize revolution, unlike Asia's rice and wheat revolutions, has yet to fulfill its early promise. Farmers have not seen sustained growth in maize yields. Although maize yields grew at an average rate of 7.1% annually over 1963–1991, growth rates declined significantly after the mid-1970s (Table 1.1). As Kenya begins to design a second generation of maize technologies, the differences between the Asian and African contexts – and the challenges for research – become more pronounced.

One great difference is water. Sustained yield increases in Kenya must be achieved under rainfed conditions, whereas large areas of Asia are irrigated. In Kenya's maize-growing areas, rainfall fluctuates significantly from year to year as well as from district to district in any particular year. This variability makes input use more risky and significantly complicates plant breeding (see the discussion in Chapters 4, 5, and 6).

Second, the gap between yields achieved on experiment stations and in farmers' fields remains large. Maize is still a relatively new crop in the country, and farmers have not learned the best ways of managing it. Input use, especially fertilizer use, remains well below recommended levels, largely because of the significant capital constraints faced by smallholder producers (see Chapter 8).

A third factor complicating the development of second-generation technologies is continued high population growth in rural areas. This population growth leads to further reductions in farm sizes, shifts maize production into increasingly marginal areas, and increases agricultural intensification. Intercropping of beans and maize has grown rapidly, along with the area of maize that is double-cropped; diversification into higher value crops has increased; traditional methods of maintaining soil fertility have been lost; and pest and disease attacks are on the rise (see Chapters 4 and 5).

Finally, the rapid – and not well articulated – liberalization of the Kenyan agricultural economy has created a more uncertain environment for farmers and for the private sector, even though the private sector is expected to make the investments required to fill the void left by the public sector's withdrawal from certain activities. All of these circumstances raise major questions for the Kenya Agricultural Research Institute (KARI), which must determine how and where maize research is most likely to maintain yield growth into the twenty-first century.

Defining and Organizing the Research Agenda for Maize in Kenya

Compared with maize research in many countries of sub-Saharan Africa, Kenya's maize research record stands out as a clear success. One purpose of this book is to review the basis for that success and to assess how widely it was shared across different maize-growing areas in Kenya. A second purpose of this book is to explore future directions for maize research. Should the research program follow the path that yielded its earlier successes, or are major modifications necessary to achieve future yield gains? Such course corrections are not easy. Changes in research direction can aggravate institutional frictions. Moreover, it has been difficult to argue convincingly that future success will depend on a different kind of research strategy, even though it is possible to make the case that the 'easy' breeding gains may have been made already. The MDBP thus was initiated at a critical time in the evolution of maize research in Kenya.

Breeding has been central to KARI's maize research program and to the success of maize technology in general in Africa. The search for materials possessing a higher inherent yielding ability across the diverse maize-growing environments in Kenya has dominated breeding strategy, and dealing with agroclimatic diversity remains crucial for developing a sound maize breeding strategy in Kenya. Yet prior to the MDBP, researchers lacked a framework for fully assessing the number and representativeness of sites that the breeding program used for selecting germplasm. The characterization of maize agroclimatic zones was a principal element of the MDBP. Because spatial diversity is now easily characterized with databases and geographical information systems (GISs), building an interactive database became the organizational basis for the MDBP. As subsequent chapters of this book will show, the KARI maize program needs to expand its number of target ecologies further, and refining the characterization of these ecologies should be a continuing research objective of the breeding program itself (see Chapters 2 and 4).

Those responsible for planning Kenya's future maize research strategy must address two central questions, which the MDBP was designed to help answer. First, should the balance of objectives change within the breeding program itself? Second, what should be the relative balance between breeding and crop management research? With regard to the first question, a dominant concern is whether to move away from the breeding objective of enhancing inherent yielding ability, towards breeding to overcome the constraints that inhibit the realization of yield potential. Of course breeding is not the only avenue for ameliorating such constraints, but it is usually a critical element in dealing with most pest and disease problems. Pest and disease resistance has in general been the principal focus of second-generation breeding in rice and wheat. To make the transition from first- to second-generation breeding objectives, KARI must assess pest and disease constraints and expand expertise in pathology and entomology. Pest and disease complexes will vary by ecology, and the Maize Data Base provides a framework for evaluating these complexes in Kenya. The results presented here rely only on farmer assessment, and thus they are a first approximation of the relative importance of different maize pests and diseases. Nevertheless, the chapter on *Striga* (Chapter 9) demonstrates the capacity of the Maize Data Base to delineate pest and disease problems and help clarify the factors governing their spread. Most important, the framework now exists to deepen the pest and disease assessment and potentially link it to a monitoring system.

In determining the relative balance between breeding and crop management research, two important considerations are the declining returns to breeding purely for higher yield and the higher costs of breeding for resistance to diseases and pests. Farmers, especially smallholders, could increase yields significantly with changes in crop management as well as wider adoption of modern varieties (Chapters 7 and 8). This book analyzes why farmers are not achieving higher yields and whether this problem should be resolved by improving the efficiency of farmer support systems (from extension through input delivery to policy) or by conducting more focused research on crop management (Chapters 7, 8, and 10) in addition to breeding research. The analysis reveals that it is not only possible, but necessary, to achieve substantial yield gains through research on improved soil fertility management, before additional investment is made in breeding. The most promising means of improving soil fertility – by better targeting of fertilizers, by adjustments in fertilizer marketing, or by expanding the range of soil management interventions, from improved management of organic resources to agroforestry – are still to be determined. The MDBP took a preliminary step toward answering that question by integrating the Kenya maize database with the database developed by the Fertilizer Use Recommendation Project (FURP), which contains data from fertilizer response trials conducted at more than 70 sites in Kenya (Chapter 8). The GIS provides a spatial framework to begin analyzing some of these issues, and further demonstrates the robustness of the Maize Data Base as an organic research tool.

The Institutional Setting for the Maize Data Base

The strategic research choices we have just listed, and which are explored in this book, will be made within a particular institutional and organizational context which is useful to describe at the outset. It is the great agroclimatic diversity of Kenya that has determined KARI's organizational structure. The extent of this diversity is reflected in the fact that KARI is the third largest agricultural research system in sub-Saharan Africa, surpassed only by South Africa and Nigeria (Pardey *et al.* 1995). The Institute is best described as a distributed system formed of a matrix of 24 research stations, 13 with national research mandates, seven with regional research mandates, and four with both national and regional mandates. Each of the 11 regional centers has responsibility for conducting the more applied and adaptive research for a particular, contiguous region of the country. The national programs in turn are organized into commodity and factor research programs and are based in designated research stations, but they coordinate research in many of the other research centers. The system seeks to balance the development of a critical mass of scientists to address more strategic research within national programs with the applied and adaptive research on location-specific problems within the regional research programs. The advantage of such a system is that economies of scale are realized in the research process at the same time that effective decentralization allows fine tuning of research to specific conditions across a wide spectrum of agroclimatic and socioeconomic diversity. The disadvantage is that coordination and transactions costs are very high.

The national maize program, which focuses principally on breeding, is based in one center, Kitale, but it must collaborate with other centers to set up selection and evaluation plots, agronomic trials, and other kinds of trials. Still other maize research activities are located within the national crop protection program, the national biological control program, the national soil fertility and plant nutrition program, and the socioeconomics division of KARI. The maize research strategy is therefore developed within a complex institutional framework, in which coordination and interprogram links are key to making the strategy operational. More radical changes in strategy are more difficult to make within such a framework. However, the Kenya maize database can bring more information, in a more systematized form, into the process of developing a research strategy, and it can also provide valuable support for organizing the work once research decisions have been made.

The planning of maize research is best seen as taking place within a *technology cycle*, which consists of designing research options and setting priorities among them; testing and adapting components of the technology at the farm level; evaluating the adoption and diffusion of the technology; and, finally, assessing the technology's impact. Each successive stage feeds back into the technology design process, which allows research strategies to be refined and better targeted (Lynam 1994). The success of this process depends on a systematic flow of information from the field and on structuring that information within an analytical framework. The approach taken by the MDBP represents a pilot effort by KARI to develop such an information system. There was a conscious focus on effectively integrating the system within KARI's organizational structure. A summary of the principal design and institutional issues involved in that process follows.

Database design

Two design components were essential to achieving the MDBP's objective of characterizing maize cropping systems in Kenya so that researchers could develop interventions with the best potential to raise productivity. These were (i) a national sampling frame for maize in Kenya and (ii) a standardized, minimum data set for characterizing maize systems. Because the database was intended to characterize the spatial diversity in maize systems as well as to reflect the underlying determinants of national maize production, the maize sampling frame needed to combine elements of both spatial- and population-based sampling frames. Spatial stratification of maize production areas by agroclimatic attributes produced a zonation scheme for maize adaptation. A population density map was overlaid on this, producing climate–population strata. A maize density map was then overlaid on this to determine sampling fractions for each climate–population stratum, so that each stratum had a weight reflecting maize production and rural population, a sufficiently close proxy for maize producers (Chapters 2 and 3).

The population-based national sampling frame was overlaid on the climate–population stratification, and a random spatial search was applied to select sites for a national survey of maize producers. A design issue that arose in constructing the database was the allocation of the survey resources between numbers of sites versus numbers of farmers per site. No prior studies were available to guide this determination, but an *ex post* assessment suggested that additional information could be gained by sampling more of the spatial diversity (sites) than by sampling more of the socioeconomic diversity (farmers) within sites. Many of KARI's regional centers may require a denser data set for their mandate areas. The maize database provides either a sampling frame within which additional surveys can be done and/or a weighting structure for the analysis of additional survey points. It should offer a robust framework for systematically evaluating the maize technology cycle over time at the national as well as regional and more local levels.

Institutionalizing a maize information system

A database system is only as good as the process and institutional structures created to make it operative. The plethora of very different surveys done to initiate past fieldwork could not be used for systematically constructing a knowledge base of maize cropping systems in Kenya, and the maize database was developed to provide a national assessment of maize technology utilization and a framework to guide and integrate future survey work. Future surveys can be designed on the basis of current assessments of what is known about maize production – the state of the database at that point in time – and will in turn serve to update the database. This objective is more easily achieved within a centralized maize program; a decentralized context such as KARI presents more challenges.

The first challenge to making the database work in a decentralized setting was to ensure that the links to data collection were strong. Initially a standardized, minimum data set characterizing maize cropping systems was constructed, based on the national survey of 1400 farmers that was designed using the sampling frame described above. Maize researchers in each of the principal research stations were involved in constructing and applying the survey instrument, and a team of researchers within each research center was trained to use the database software and GIS. Each team analyzed the data from its particular mandate area. The team was thus familiar with what was in the database as well as the means to interrogate the database.

A subsequent challenge will be to ensure that the database is linked to future data collection efforts. Such efforts could take three principal forms: increasing the density of the coverage by expanding the number of sites and farmers; increasing the depth of the survey by, for example, soil tests of farmers' plots or monitoring of disease and pest severity; or conducting the original survey again to develop a longitudinal data set. It is hoped that the database will expand in all three directions and that it will continue to rely on the network of survey teams that has been developed in each of the research stations. Whether this kind of work will eventually evolve into more of a monitoring system depends on decisions about the future objectives of the Kenya maize database.

Another challenge in successfully implementing the database within KARI is to ensure that the coordination and maintenance of the database are centralized functions. Coordination or backstopping will be needed for sampling, survey design, and the development of specialized protocols, such as a protocol for measuring the severity of stem borer attack. Centralization would help develop a suite of survey instruments that would be efficient in collecting the required data and would produce a more standardized set of data. Moreover, a centralized unit is required to maintain and update the database. Such a unit would ideally have capacity in GIS, so new layers of data could be added to improve the analytical power of the system (e.g., input and output price isolines for the country). The MDBP has generated a discussion of these issues within KARI, but they have not been resolved. Data management and data access remain emerging issues within KARI, as in most African national research systems.

A final challenge is to link the database to analytical and planning capacity in KARI. It seems logical for both regional and national maize research planning to be based on the same data set, but this was not possible before the maize database was developed. In fact, before the MDBP, maize research planning was primarily done regionally, as there was no basis for a national planning exercise. The maize database augments KARI's priority setting capacity and research planning within the national maize program. However, analytical capacity, especially in economics, is a limited resource in KARI, and the power of the database will be limited as well unless the information it contains can be analyzed effectively by more persons, both within and outside KARI. This will require a policy on data distribution and sharing, but such a policy has not been developed.

Targeting within the technology cycle

The maize database was designed to provide easy access to critical information for planning research and developing technology. The national sampling frame, the explicit spatial coverage and links to GIS, and the framework for adding to the database all were designed so that technologies could be targeted better and tracked more easily through the technology cycle. The following paragraphs briefly review some of the applications of the database.

At the stage of research planning and technology design, the most obvious impact has been improved stratification of target environments for the breeding program (Chapters 2 and 4). In particular, an additional varietal development project was added for the moist transitional zone. This 'new' zone, identified during the development of the database, is a somewhat warmer area characterized by bimodal rainfall, where farmers are increasingly moving to double-cropping although their maize varieties do not mature rapidly enough to be ideal for the new cropping systems. Time to maturity is a key trait for developing appropriate maize varieties in a country like Kenya, as this trait determines a variety's adaptation to rainfall risk in the semiarid regions, susceptibility to ear rot (if a variety does not mature until the next rains begin), or a reduction in yield potential if a variety matures in too short a time. The time to maturity most appropriate for maize germplasm in a given set of circumstances can be determined quite accurately using crop modeling (Keating *et al.* 1994), and an effort is under way to link the CERES-MAIZE crop model to the Kenya maize database.

As observed earlier, the breeding program will require more information on pest and disease distribution if it is to shift its emphasis from breeding for yield potential to breeding for the spectrum of constraints in well-defined agroecologies. This shift in emphasis will require plant health assessment protocols to be developed and then applied in field-based research, or it will require the institution of a coordinated monitoring system. Shifting the emphasis to pest and disease resistance can increase the costs of breeding substantially, and a better assessment of actual yield losses from pests and diseases would aid considerably in setting future breeding strategies (Chapters 4 and 6).

Since the advent of the maize database, KARI has significantly strengthened its adaptive research capacity within the regional research centers. This *in situ* field research capacity can take advantage of the database for developing prototype interventions and can also add information to the database as on-farm research proceeds. It is too early to evaluate the application of the database in selecting sites and prototype technologies for on-farm testing, even though staff in KARI's regional research centers have been trained to use the database to do so. Several factors, however, limit the utility of the database in its present form: the low resolution and low density of sites for database applications at a regional level; the minimal information on other components in the farming system; and movement toward participatory research approaches in defining technology interventions. The eventual link to the modeling component and a higher resolution database could add to the power of the database for use in adaptive research, particularly in conjunction with participatory research methods. On the other hand, on-farm adaptive research could enhance the power of the database because of the potential for generating an in-depth evaluation of constraints on maize yield. There are, however, trade-offs here in relation to how standardized the data should be, how many resources should be devoted to monitoring constraints, and how much researcher management should go into the trial. Information generated by on-farm research is not often applicable in other settings, yet the more standardization, the more utility the information will have in improving the understanding of the limits of current maize technologies.

Finally, as most of this book demonstrates, the maize database provides a systematic basis for evaluating the adoption and impact of technology (Chapters 7 and 8). How often this kind of evaluation should be done is probably best gauged by results from adaptive research, but for these studies to be most effective, they are probably best done as a coordinated, national study. Independent adoption studies undertaken by the regional centers will have some value, but will certainly not yield the insights that a national study would. In this sense, the maize database provides something of a model on how such a study might be carried out, but with the added dimension that *ex post* analysis feeds directly into *ex ante* research planning (Chapter 6).

Conclusion

Planning maize research in Kenya demonstrates in many ways the challenges to producing improved agricultural technologies in Africa. Maize researchers must confront a significant agroclimatic and socioeconomic heterogeneity, a smallholder farming population operating on a very constrained resource base, and a maize sector undergoing significant structural change. Designing improved maize technologies in such an environment puts a premium on information that characterizes the maize target area, the variation in maize producers and cropping systems within the target area, and changes in those systems over time. And yet, such information in Kenya, and Africa in general, is very limited, very costly to collect, and difficult to assemble in a form useful for planning agricultural research. The Kenya Maize Data Base Project was established as a prototype for the consistent development of such an information base for agricultural research planning, monitoring, and impact assessment. It utilizes new database and GIS technologies, a sampling frame particularly designed to meet the needs of agricultural research planning, typical farm survey instruments, and the extensive field research capacity which is one of the strengths of the KARI system. The succeeding chapters give equal weight to how such an information system is developed and institutionalized within KARI and to the results of the analysis of that data, particularly focusing on the current state of maize technology adoption and drawing the implications for the next generation of maize technologies that is required to maintain the needed growth in maize productivity in Kenya.

References

Eicher, C. (1995) Zimbabwe's maize-based green revolution: Preconditions for replication. *World Development* 23(5), 805–818.

Gerhart, J. (1975) *The Diffusion of Hybrid Maize in Western Kenya.* Research Report for CIMMYT. International Maize and Wheat Improvement Center (CIMMYT), Mexico City. Mimeo.

Keating, B., Wafula, B., Watiki, J., and Karanja, D. (1994) Dealing with climatic risk in agricultural research: A case study modelling maize in semi-arid Kenya. In: Craswell, E. and Simpson, J. (eds) *Soil Fertility and Climatic Constraints in Dryland Agriculture.* ACIAR Proceedings No. 54. ACIAR, Canberra, Australia.

Lynam, J. (1994) Integrating research planning, priority-setting, and impact evaluation within the CGIAR. In: Collinson, M. and Platais, K. (eds) *Social Science in the CGIAR.* CGIAR Study Paper No. 28. World Bank, Washington, DC.

Ministry of Planning and National Development. Various years. *Statistical Abstracts.* Nairobi, Kenya: Ministry of Planning and National Development.

Miracle, M. (1966) *Maize in Tropical Africa.* University of Wisconsin Press, Madison, Wisconsin.

Pardey, P., Roseboom, G., and Bietman, N. (1995) *Agricultural Research in Africa: Three Decades of Development.* Briefing Paper 19. International Service for National Agricultural Research (ISNAR), The Hague, The Netherlands.

Endnote

1 First by the Maize Marketing Board and subsequently by its successor organization, the National Cereals and Produce Board.

2 Classifying Maize Production Zones in Kenya through Multivariate Cluster Analysis

JOHN D. CORBETT

The Kenya Maize Data Base Project analyzed mean, minimum, and maximum monthly air temperatures compiled from 81 meteorological stations and mean monthly precipitation data from 979 stations to differentiate among maize production environments. These data sets were augmented by spatial interpolation to fill all valid ground grid cells of the 3 arc-minute digital elevation model for constructing climate surfaces. Cells with similar climatic properties were grouped together through a clustering process. Classification criteria consistent with those used by the Kenyan national maize research program were used to delineate agroclimatic zones for maize production. Eight zones were identified: the dry and moist lowland tropics; dry and moist midaltitude zones; dry and moist highland tropics; cool highland tropics; and a zone characterized by extreme water stress.

Because an accurate definition of production environments is crucial for targeting maize germplasm to the environment where it can best express its potential yield, Kenya has organized its maize research to serve several broadly defined maize production environments, referred to as 'adaptation zones.' However, these zones were defined using only parameters of altitude and moisture availability (Gebrekidan *et al.* 1992), and their usefulness was further limited because researchers lacked an adequate 'spatial description' or map of the boundaries of the zones in digital form. The two schemes of agroclimatic zonation available for Kenya (Sombroek *et al.* 1982; Jaetzold and Schmidt 1983) were not detailed enough to apply to individual crops and suffered from other limitations as well. Fortunately, the emergence of GIS has made it possible to delineate agroclimatic zones with greater precision, especially by allowing many 'layers' of spatially referenced data (including survey data) to be integrated into one digital database.

As stated in the previous chapter, one of the MDBP's objectives was to use GIS and spatial analysis to improve the characterization of Kenya's agroclimatic diversity and refine the definition of maize production environments. The first task of the MDBP was to reproduce KARI's present classification of maize environments in digital form. This chapter discusses why digital, GIS-based approaches are replacing traditional cartographic methods for classifying agroclimates and describes how the MDBP used GIS techniques to construct maize-specific agroclimatic zones for Kenya. The chapter ends with a discussion of the results and implications of that spatial analysis for future research.

Limitations of Traditional Approaches for Delineating Agroclimatic Zones

The characterization of agricultural production environments has been the subject of much research (see, for example, Koppen and Geiger 1936; Thornthwaite 1948; and FAO 1981) to integrate available data with expert opinion and obtain an interpretation of the resource base for agricultural development. Traditional models of agroecological characterization have relied on cartographic reproduction of static zones based on fixed crop–environment relationships, but these approaches have several major limitations:

1. They define a single set of zones applicable to all crops. The zones are too general to reflect the range of sensitivity of most crops to environmental variables.

2. Because they use only a limited, fixed number of climatic parameters, traditional approaches are unable to assess how these parameters might interact through the growing season (e.g., they cannot assess the implications of different combinations of temperature and precipitation from planting to harvest). Different combinations of variables and boundary conditions are needed to assess the suitability of a given climate for each crop and crop genotype.

3. The use of discrete ranges of descriptive variables often leads to serious errors. If lowland areas occur only between 0 and 1000 m, lowland-type environments at slightly higher elevations are, by definition, excluded. Static zoning cannot accurately capture the potential distribution of a crop's genotypes and ecosystems.

4. Different spatial and temporal data cannot be integrated. Traditional systems cannot take advantage of new data of various resolutions as they become available.

5. Static classification ignores the specific requirements of different genotypes of a crop species. Maize varieties, for instance, may take from around 80 to more than 300 days to mature, depending on the individual genotype and the location where it is grown. Although traditional agroecological classifications attempt to take account of the growing-season temperatures and moisture required by specific crops, the variation among genotypes is either not recognized or oversimplified. Some maize genotypes are quite sensitive to drought stress at flowering, but breeders can attempt to reduce the anthesis–silking interval of a particular genotype to improve its drought tolerance. A more flexible method for defining production environments could identify specific areas where drought stress becomes critical at flowering. Research priorities would be influenced depending on the extent of the stress zone. The ability to target areas suitable for a particular type of germplasm both facilitates priority setting for research and increases the efficiency of research.

6. Finally, an assessment of risk (timing, frequency, and intensity of abiotic and biotic constraints) would be extremely useful in characterizing cropping environments. In semiarid areas it is climatic variability, not climate averages, that characterizes farming systems. Classification systems in Kenya used discrete ranges of descriptive climatic variables and could not integrate the timing and frequency of biotic and abiotic phenomena into their agroecological classifications.

A New Approach to Zonation

The approach employed by the MDBP used a digital framework to characterize agroclimatic zones, and it possesses several advantages that enable it to overcome the limitations of the static systems outlined earlier:

1. It can be specific to a given crop, type of germplasm, and/or ecosystem.
2. It has the capacity to integrate various combinations of descriptive variables (e.g., it can deal with multiple constraints to producing a certain crop).
3. It can integrate various spatial and temporal scales.
4. It can be employed for risk assessment.

Among these advantages, the most important is the ability to delineate different zones using the same digital characterization database. This makes it possible to evaluate a given zonal classification based on germplasm or crop traits and not on the interpolated (traditionally contour maps) data. The ability to reproduce boundaries digitally also allows for subsequent improvements in the classification as researchers gain a better understanding of crop–environment interactions or acquire more data. At first, similar climates may be grouped using criteria specific to a crop. The diversity among genotypes of the crop is taken into account because the climate classification uses characteristics of adaptation (e.g., disease susceptibility) that are specific to different types of germplasm. Next, these crop characteristics can be combined with their constraints (using different sets of variables) to result in a more precise classification. For example, just when the temperature range becomes optimal for a certain disease to develop, a specific genotype may enter a growth stage at which it is susceptible to that very disease. The defining variables for the crop and the disease can be classified separately but combined to delineate the genotype's target areas based on the crop–climate characteristics and the disease's epidemiology (as related to climatic variables). The result is a digital surface that is targeted to a *specific* genotype and can be reproduced graphically. The kind of classification just described can be taken even further by integrating agrotechnical data with socioeconomic information to evaluate the implications of agricultural policies. A final advantages is that the results of the analysis – even the data themselves – are easy to transfer.

Though the MDBP did not include risk assessment in its characterization of environments, this possibility should be examined in the future. Data on the timing of specific conditions can greatly improve the targeting of germplasm to specific areas and thus offer better information for setting research priorities. In some tropical areas such as eastern Kenya, the rainy season can be thought of as having a 'switch': either the rains arrive or they do not. The probability of good rains has a spatial component (in Kenya, it is related to elevation). Thus the targeting of germplasm (and advice related to the production system) can be enhanced through an evaluation of the variability in rainfall across space.

Construction of Digital Climate Surfaces

Accurate and robust geo-referenced digital climate surfaces are an essential requirement for a dynamic approach to agroclimatic classification. Hutchinson *et al.* (1992) favor initially using mean monthly data to describe climatic variation. Mean monthly climate values show significant variation at both the microscale and mesoscale. At a still larger scale, temperature and precipitation are affected by elevation. In tropical areas, temperatures decrease (roughly) by about 6°C for every 1000 m rise in elevation – a condition referred to as the 'temperature lapse rate.' The effect of elevation on precipitation is less systematic but still significant. These dependencies can be applied when interpolating monthly mean climate values to areas between meteorological stations. The Laplacian method (thin plate smoothing splines) (Wahba and Wendelberger 1980) is used to incorporate the independent variable, elevation, with the two usual spatial variables (latitude and longitude).

The degree of data smoothing imposed by each fitted surface is determined objectively by minimizing the generalized cross validation (GCV) error. The predictive error of the fitted surface is minimized by implicitly withholding each data point in turn in order to validate the fitted surface. The variance of the errors associated with each data value is then estimated by analogy with linear regression (Wahba 1990), from which an estimate of the 'true' standard error of the fitted values can be calculated.

Monthly climate surfaces are constructed from period normal data (minimum and maximum temperature, precipitation, evaporation, radiation, and humidity). Thin plate smoothing splines produce *fitted surfaces.* The fitted surfaces represent climate variables (e.g., mean monthly maximum and minimum temperatures) as trivariate spline functions of longitude, latitude, and elevation. This procedure takes advantage of the strong dependence of climate variables, especially temperature, on elevation, but allows the size of this dependence (e.g., the temperature lapse rate) to vary over space and time (Hutchinson 1991b).[1] In this case, the temporal variation is between months, meaning that the lapse rate (the temperate/elevation relationship) is calculated for each month. Incorporating the elevation dependence, in addition to the usual dependence on longitude and latitude, helps overcome the scarcity of data. Elevation data from topographic maps become the basis for interpolation between locations where climate variables are measured. Standard interpolation packages rarely provide for more than bivariate interpolation (Hutchinson 1991a).

Precipitation and evapotranspiration also show significant dependencies on elevation, although these dependencies are less systematic than that of temperature and elevation. For precipitation, the error is strongly related to the density of stations from which data are taken: more stations will improve the surface for this parameter, which is highly spatially variable. Station density is even more important on complex terrain, as is a detailed digital elevation model.

Climate data are scarce and/or of widely varying quality for many areas in Kenya. The results of the spline interpolation provide feedback on the meteorological station database by providing the deviation of values at each station relative to the fitted surface values. Each station can be evaluated in the context of the neighboring stations, and the reliability of the measured value can then be assessed.

The Data and Zonation Procedures for Kenya

The data sets

Several data sets were used to develop the maize agroecological zones for Kenya. In 1992, the Australian Centre for Resource and Environmental Studies released a digital elevation model (DEM) for all of Africa. The Australian model, developed from many sources of topographic data, was the source of the Kenya DEM. The original topographic base maps were of a nominal scale of 1:1,000,000. The Kenya DEM has a resolution of 3 arc-minutes or about 5.5 km at the equator.

To construct the temperature surfaces, long-term monthly minimum and maximum temperature data[2] were obtained from 81 meteorological stations for Kenya (Jones 1988). Year-to-year mean monthly temperatures generally do not vary greatly in the tropics, and Woodhead (1968) and others have shown that temperature values measured for Kenya are consistent with this pattern of small interannual variability. Also, because temperature is strongly dependent on elevation, a less dense network of meteorological stations is needed to gather temperature data than precipitation data. On the other hand, Kenya's precipitation network is quite extensive, especially in the arable areas. Data from 979 stations were used to interpolate the precipitation surface (Jones 1988).

The maize zonation methods

The data sets were processed using a three-stage sequence of agroclimatic classification procedures to develop the maize-specific agroclimatic zones for Kenya.

1. Spatial interpolation

Climate surfaces were constructed using the techniques of spatial interpolation reviewed earlier. The fit of the surfaces is improved if abrupt boundaries can be avoided, so stations across the border from Kenya in Tanzania and Uganda were included to create a more continuous surface.

The data sets proved quite robust. Initial spline calculations identified only a few stations that needed correcting for latitude, longitude, and/or elevation. The errors, measured by the square root of the GCV term for the 12 monthly minimum and maximum temperature and monthly precipitation climate surfaces, are given in Table 2.1. The square root of the GCV is analogous to the mean squared deviation of the measured points from the fitted surface. Note also that the square root of the GCV is a conservative indication of the accuracy of the interpolated surface.

To demonstrate the deviation of the fitted surface from individual points, the four precipitation stations with the largest mean deviation over the 12 months and the station with the one-hundredth largest deviation are provided in Table 2.2. From the table it can be seen that 879 precipitation stations in the data set have a root mean square error of less than 31 mm over 12 months.

2. Multivariate cluster analysis

Once the climate surfaces were constructed, grid cells possessing similar climatic attributes were grouped together using hierarchical cluster analysis – specifically, Ward's minimum-variance clustering method (Ward 1963; SAS Institute 1991). The algorithm begins by computing a matrix of squared Euclidean distances between every possible pair of grid cells for each variable.

Table 2.1. Square root of the generalized cross validation (root GCV) error term for mean monthly climate attributes in Kenya.

	Root GCV		
Root GCV	Min. temp. (°C)	Max. temp. (°C)	Precipitation (mm)
January	1.08	1.00	13.4
February	1.23	1.02	15.3
March	0.97	0.97	23.5
April	0.76	0.93	45.8
May	0.80	0.80	45.1
June	1.04	0.79	21.5
July	0.95	0.68	20.9
August	0.96	0.61	20.4
September	1.05	0.76	18.6
October	0.89	0.93	26.2
November	0.82	0.98	39.6
December	0.99	1.32	23.6

Source: MDBP.

Table 2.2. Ranked root mean squared error for precipitation (mean over 12 months) for five stations used to create climate surfaces.

Rank	Station	Precipitation (mm)	Longitude (°)	Latitude (°)	Elevation (m)
1	Lyanmungu, Tanzania	102	37.24 E	−3.23	1250
2	Gatare Forest, Kenya	86	36.76 E	−0.71	2590
3	Mawingo Estate, Tanzania	83	37.33 E	−3.23	760
4	Maua Nyembene, Kenya	82	37.94 E	0.24	1620
–					
100	Sotik Chelalungu, Kenya	31	35.13 E	−0.91	1770

Source: MDBP.

Each cell is initially considered a separate cluster. Based on the analysis, clusters with similar characteristics are merged in a stepwise fashion. Ward's minimum-variance method computes distances between clusters, added over the variables. Within-cluster sums of squares, divided by the total sum of squares, give estimates of proportions of the variance. At each step, the within-cluster sums of squares are minimized by merging the two most similar clusters. The analysis was stopped at 50 clusters, a limit chosen subjectively as a 'reasonable' number of clusters on which to apply maize experts' criteria for separating adaptation zones (stage 3).

Three variables – long-term mean minimum temperature, mean maximum temperature, and mean precipitation – were used for each of the six months in the growing season (March–August), contributing a total of 18 variables to the cluster analysis. By using data for each month rather than using average total values for the season, the cluster analysis captures important monthly variations to which maize growth and development may be very sensitive. This was the main reason for using cluster analysis as an intermediate stage between constructing the climate surfaces (stage 1) and deriving the adaptation zones (stage 3). If the zones had been interpreted directly from the climate surface data, grid cells with similar levels of precipitation for the entire season would have been grouped together, even if those cells had significant month-to-month variations in rainfall patterns. Because plants are more sensitive to the timing of rainfall and temperature levels at some stages of growth than at others, it is crucial to capture at least the variation between months, if not within months, to arrive at a more accurate differentiation of agroclimates. This kind of time-series differentiation is impossible without the climate surface database. Such databases afford researchers immense flexibility. More variables, such as elevation, extent of vulnerability to erosion, and variance measures, can be readily added to the current set of parameters to distinguish adaptation zones based on additional criteria.

Elevation data were added to the 18 climatic variables described earlier. The use of elevation in the clustering is somewhat controversial, because the temperature layers use the same elevation data in combination with the statistical surfaces from the splines. The decision to include elevation related primarily to the desire to offer these maize-specific agroclimatic zones as a bridge between the traditional zonation schemes currently used in Kenya and a more dynamic, iterative methodology. Most researchers in Kenya are quite familiar with elevation criteria, and including elevation in the clustering was a mechanism for drawing on their prior knowledge. Another advantage of a clustering with climate and elevation data is that environments that do not belong together will no longer be grouped together just because they fall on the same side of an arbitrary and discrete boundary (e.g., a contour line for a given elevation). A midaltitude environment could occur below 1000 m just as a lowland environment might occur above 1000 m.

3. Delineation of maize agroclimatic zones

The results of the cluster analysis were interpreted into maize adaptation zones based on the knowledge and experience of maize scientists at KARI. These zones correspond to broad classes of maize germplasm (e.g., lowland tropical maize germplasm and midaltitude maize germplasm) as well as groups of constraints (e.g., diseases and pests). Plant breeders seek to improve maize yields by developing materials capable of overcoming the suite of constraints found in a target region. Lowland materials, for example, need no resistance to disease caused by *Exserohilum turcicum*, though in some areas they must be resistant to downy mildew disease. Highland maize materials are distinguished by their ability to tolerate cold and cool temperatures and the diseases that may be present because of these biophysical conditions.

Maize research programs typically divide their work into categories that reflect broad zones of adaptation, and the Kenyan program is no exception. Based on elevation, three adaptation zones were defined for maize germplasm in Kenya: the lowland tropics at less than 1000 m above sea level; the midaltitude tropics at 1000–1800 m above sea level; and the highland tropics at more than 1800 m above sea level. The midaltitude tropical zone was then divided into dry (semiarid) and moist zones. Since the early 1960s, maize research by KARI has been organized to serve these four adaptation zones. Eventually a fifth research program was added to serve very high altitude areas exceeding 2300 m above sea level (Gebrekidan *et al.* 1992).

The new classification by the MDBP added more moisture and temperature regimes within each of the three major elevation zones. Maize scientists at KARI and CIMMYT defined rainfall and temperature ranges that would represent meaningful variations for maize adaptation and lead to a more comprehensive characterization of maize production environments. These multivariate criteria (listed in Table 2.3) were applied to the final clusters of climate surfaces to define maize-specific agroclimatic zones. The resulting classification essentially reproduced the maize zones used by KARI. It added subcategories defined by moisture availability within each elevation range and a zone characterized by extreme water stress (Geographical Data Table 1, Plate 1).

Table 2.3. Maize scientists' criteria for distinguishing among adaptation zones, Kenya Agricultural Research Institute.

Zone	Elevation (m)	Temperature (°C)		
		Minimum	Maximum	Mean
Lowland tropics	< 1000	22	32	28
Midaltitude tropics	1000–1800	17	32	25
Highland tropics	> 1800	7	24	16
	Total precipitation (mm)			
Dry	< 500			
Moist	500–1000			
Wet	> 1000			

Source: MDBP.

Results and Implications of the Spatial Characterization of Maize Agroclimates

The zoning exercise described in the previous section made it possible to produce a map of Kenya's maize production environments for maize researchers at KARI. For the first time, researchers had a spatial representation of the maize adaptation zones that KARI used to organize its maize research. The map reflects only a single evaluation of maize production environments in Kenya, and it can be revised and expanded in any number of ways to take other information into account. The database is designed (and available for) more regionally specific analyses. Researchers are aware that the primary growing season in the dry eastern lands of Kitui and Machakos is associated with the November rains, just as an important maize growing season in the Lake Victoria basin takes advantage of the January and February rains in that region. The maize adaptation zones might be improved by building a database in which each cell is given the number associated with the actual optimum month of planting, which depends on when the rains begin. The cluster analysis would then use this month as month one and the following months as the growing season. Future efforts to improve the maize zonation scheme might begin here.

As stated earlier, the spatial analysis described in this chapter was the starting point of a wider exploration of the parameters of maize production in Kenya. The next chapter describes how this analysis was extended to obtain a more accurate and objective approximation of the relative size of each maize production zone, which was crucial for designing a national survey of maize farmers.

References

FAO (Food and Agriculture Organization of the United Nations). (1981) *Report on the Agro-Ecological Zones Project*. Vol. 1 of *Methodology and Results for Africa*. FAO, Rome, Italy.

Gebrekidan, B., Wafula, B., and Njoroge. K. (1992) Agroecological zoning in relation to maize research priorities in Kenya. In: *KARI: Review of the National Maize Research Program. Proceedings of a Workshop Held November 19–23, in Kakamega*. Kenya Agricultural Research Institute (KARI), Nairobi, Kenya.

Hutchinson, M.F. (1991a) The application of thin plate smoothing splines to continent-wide data assimilation. In: Jasper, J.D. (ed.) *Data Assimilation Systems*. BMRC Research Report No. 27. Bureau of Meteorology, Melbourne, Australia, pp. 104–113.

Hutchinson, M.F. (1991b) Estimating the spatial distribution of monthly mean and extreme temperatures across China. *Conference on Agricultural Meteorology*. Bureau of Meteorology, Melbourne, Australia, pp. 319–322.

Hutchinson, M.F., Nix, H.A. and McMahan, J.P. (1992) Climate constraints on cropping systems. In: Pearson, C.J. (ed.) *Ecosystems of the World: Field Crops Ecosystems*. Elsevier, Amsterdam, The Netherlands, Chapter 3.

Jaetzold, R., and Schmidt, H. (1983) *Farm Management Handbook of Kenya: Natural Conditions and Farm Management Information*. Ministry of Agriculture, Nairobi, Kenya.

Jones, P.G. (1988) *CLIMATE, Version 3.5. World Tropical Climate Database*. Agroecological Studies Unit, Centro Internacional de Agricultura Tropical (CIAT), Cali, Colombia.

Koppen, W., and Geiger, R. (1936) *Handbuch der Klimatologie*. Gebr. Borntraeger, Berlin, Germany.

SAS Institute. 1991. *SAS Language and Procedures, Version 6*. SAS Institute, Cary, North Carolina.

Sombroek, W.G., Braun, H.M., and van der Pouw, B.J. (1982) *Exploratory Soil Map and Agro-climatic Zone Map of Kenya*. Exploratory Soil Survey Report No. E1. Kenya Soil Survey, Kenya Agricultural Research Institute (KARI), Nairobi, Kenya.

Thornthwaite, C.W. (1948) An approach to the rational classification of climate. *Geographical Review* 38, 55–94.

Wahba, G. (1990) Spline models for observational data. *CBMS-NSF Regional Conference Series in Applied Mathematics*. Society for Industrial and Applied Mathematics, Philadelphia, Pennsylvania.

Wahba, G., and Wendelberger, J. (1980) Some new mathematical methods for variational objective weather analysis using splines and cross-validation. *Monthly Weather Review* 92, 169–176.

Ward, J.H. (1963) Hierarchical grouping to optimize an objective function. *Journal of the American Statistical Association* 58, 236–244.

Woodhead, T. (1968) *Studies of Potential Evaporation in Kenya*. Water Development Department, Ministry of Natural Resources, Nairobi, Kenya.

Endnotes

1 This technique has been used to interpolate mean monthly maximum and minimum temperatures across Australia to within a standard error of 0.5°C. It has also been

used to interpolate mean monthly precipitation to within a standard error of about 10% (Hutchinson *et al.*, 1992).

2 The data series for most stations extended for more than 60 years. Even for the few stations where the data series were shorter, the shortest series exceeded 12 years.

3 The Spatial Sampling Frame and Design for Farmer and Village Surveys

RASHID M. HASSAN, JOHN LYNAM, AND PETER OKOTH

Geographic information systems techniques were used to design a spatial sampling frame for a national survey of maize farmers in Kenya. The spatial sampling frame was then used to stratify maize-producing areas into homogeneous clusters, allocate sampling densities, and select survey sites and farmers. This survey design made it possible for the MDBP to integrate spatially referenced survey data on maize production with climatic attributes and other spatial data to develop a comprehensive digital database for maize in Kenya. Furthermore, researchers will now have access to a semipermanent sample for monitoring future changes in the diffusion of maize technologies, evolution of the maize farming environment, and changes in farming system constraints. A post-survey analysis evaluated the efficiency trade-offs between increasing the number of survey sites versus interviewing more farmers per site. Gains in sampling efficiency from adding extra survey sites would have been insignificant. Parameter estimates derived from a sample of 10–15 farmers per site were not significantly different from those obtained from the original sample of 20 farmers per site.

The previous chapter described data and methods used to produce an initial schema of Kenya's agroclimatic zones for maize production. This chapter describes how the MDBP supplemented the data used to develop the agroclimatic zones by designing and implementing farmer and village surveys, which would yield data on farmers' production practices and resource base, as well as on markets, infrastructure, and other socioeconomic factors shaping maize production systems in Kenya. Geographic information systems techniques were used to design a spatial sampling frame for the surveys. The spatial sampling frame was then used to stratify maize-producing areas into homogeneous clusters, allocate sampling densities, and select sites and farmers to survey. This survey design made it possible for the MDBP to:

- Collect spatially referenced farm-level and regional data on maize production across Kenya.
- Integrate geo-referenced survey information with climatic attributes and other spatial data into a comprehensive digital database. (Although survey information and spatial data are usually analyzed separately, geo-referencing gives researchers a means of analyzing them together.)
- Use these data to assess current adoption and performance of maize production technologies in farmers' fields; characterize maize production systems; identify productivity constraints; evaluate maize research priorities; and assess the potential for increasing productivity by improving the design of maize research and more accurately targeting improved maize production technologies to particular areas and groups of farmers. (Subsequent chapters in this book address these issues.)
- Develop a semipermanent sample for monitoring future changes in the diffusion of maize technologies, evolution of the maize farming environment, and changes in farming system constraints.

Data and Methods Used to Design the Farmer Survey

Stratified random sampling was used to design the survey of maize farmers in Kenya. Proper stratification of the sample reduces the overall variance and increases sampling efficiency, so smaller samples are required to obtain a given level of precision (Lazerwitz 1971). The crucial question, however, is which criteria and attributes to use to classify the population under study. For this study, three attributes were used to stratify Kenya's maize-producing regions into relatively homogeneous clusters and determine sampling fractions: the agroclimate, population density, and intensity of maize cultivation.

These three attributes were considered important for several reasons. The limits and potential of maize production and adaptability are largely defined by the biophysical environment in which maize is grown, so the spatial diversity in maize production environments must be characterized as a first step towards properly evaluating whether research is targeted well and towards designing relevant technology. Climatic attributes were generally considered the best criteria for differentiating maize production environments (Chapter 2; Plate 1; Pollak and Pham 1989). Given that the area and number of farmers/households likely to be affected by the introduction of new production technologies are important variables for assessing the impact and potential gains from agricultural research, both population density and the percentage area under maize were used in designing the survey.

One advantage was that data on these three attributes were available in compatible digital formats that were easy to merge. Although information on equally important biophysical attributes, such as soils, was unavailable in the desired resolution and could not be used to develop the spatial sampling frame, information collected from a series of informal surveys on cropping patterns, farmers' practices, and other socioeconomic variables was used to verify major variations between and within maize zones. These data and other secondary information were used to analyze sources of variation and to guide selection of survey sites for purposive sampling within the climate–population strata.

A description of how such information was incorporated in the spatial design of the farmer survey in Kenya's Coast Region is given later in this chapter. First, however, we will turn our attention to the components of the three data sets used to define maize adaptation zones.

The Three Data Sets

The MDBP agroclimatic classification

As described in the previous chapter, the MDBP defined eight agroclimatic zones for maize production in Kenya (Geographical Data Table 1, Plate 1). These zones formed the primary strata for purposive sampling of Kenya's maize farmers.

Population data

Population density was used for secondary stratification within the primary agroclimatic strata. The 1989 population census was used to determine population density within the maize agroclimatic zones. The number of households per sublocation was combined with a digital coverage of administrative units (sublocation map) to show the distribution of households in Kenya (CBS 1990b). To arrive at a manageable number of climate–population strata, population was grouped into four density classes: *low-density areas* (fewer than 100 persons km^{-2}); *medium-density areas* (100–400 persons km^{-2}); *high-density areas* (400–800 persons km^{-2}); and *very-high-density* areas (more than 800 persons km^{-2}). Map 1 presents the resulting population classification.

Intensity of maize cultivation

Every year, the Department of Remote Sensing and Resource Surveys (DRSRS) of the Ministry of Planning and National Development produces a map of maize cultivation intensity in Kenya using data generated by aerial photography and ground verification surveys (Otichillo and Sinange 1991). Intensity of maize cultivation is interpreted on a district basis using 5×2.5 km transects. Boundaries for the intensity of maize cultivation are then defined, dividing the country into five zones:

- A zone having *no to very low density* of maize production (maize covers less than 0.5% of the land).
- A *low-density zone* (maize covers 0.5–5% of the land).
- A *medium-density zone* (maize covers 5–10% of the land).
- A *high-density zone* (maize covers 10–20% of the land).
- A *very-high-density zone* (maize covers 20–40% of the land).

To determine sampling fractions, the maize density zone map (Plate 2) was used to compute the average maize hectareage in each climate–population class.

The Sampling Strategy and Selection Procedure

Multilayer stratification

The spatial analysis module of ARC/INFO (vector-based GIS software) was used to overlay the climate surfaces, population coverage, and intensity of maize cultivation maps (Plate 1 and 2, Map 1) so that these three layers of information were united in one database. A cross-classification of the country by agroclimate, intensity of maize cultivation, and population density strata was produced. Farmers in the resulting 32 climate–population classes (8 zones × 4 population groups) were sampled in proportion to the area of maize and population density in each class.

Table 3.1 shows the distribution of area sown to maize in 1990 for each climate–population stratum. More than 50% of the area under maize falls in the dry and moist locations of the midaltitude zone. An additional 37.9% of the maize area is found in the dry, moist, and cold locations of the high-altitude zone.

Table 3.1. Area (000 ha) under maize by agroclimatic zone and population density, Kenya, 1990.

	Population stratum					
Agroclimatic zone	Low density (< 100 pers. km^{-2})	Medium density (100–400 pers. km^{-2})	High density (400–800 pers. km^{-2})	Very high density (> 800 pers. km^{-2})	Total area (000 ha)	Percentage of total area
No maize zone	37	na	na	na	37	3.5
Dry lowland tropics	10	2.4	0.5	0.1	13	1.2
Moist lowland tropics	20.7	5.3	1.6	0.4	28	2.6
Dry midaltitude	59	44	5	2	110	10.2
Moist midaltitude	88	260	86	23	457	42.6
Dry high-altitude	23	65	9	3	100	9.3
Moist high-altitude	68	144	45	8	265	24.6
Cool high-altitude	24	18	0.7	0.3	43	4.0
Extreme water stress	13	6	2	na	21	2.0
Total area	343	544.7	149.8	36.8	1074	100.0
Percentage of total area	31.9	50.7	14.0	3.4	na	na

Source: Calculated from the DRSRS maize coverage map for 1990 (Otichillo and Sinange 1991).
na = Not applicable; pers. = persons.

About 96% of the estimated total area under maize falls within four maize cultivation intensity classes: the low-, medium-, high-, and very-high-intensity groups, which occupy about 15% of Kenya's total land area.[1] The remaining 3.5% of the maize area is located in the 0–0.5% maize intensity zone (Table 3.1), which extends across the remaining 85% of the total area of Kenya. The 0–0.5% maize intensity zone has 1 ha of maize for every 1300 ha (48 million/37,000), compared with the other intensity classes, which have an average of 1 ha of maize for every 8 ha of land (8.6 million/1.04 million). The 0–0.5% maize intensity zone also falls within the area of low population density (Plate 2). In terms of sampling efficiency, the gain from surveying an extra farmer in this zone was considered too low to justify the extra costs associated with covering areas of such low population density and maize concentration, so no farmers from this zone were sampled.

The eight agroclimatic zones considered for the survey were further subdivided into climate–population groups using population density classes (Table 3.1). Information from a series of informal surveys conducted by the MDBP revealed that population was an important factor influencing farming practices within zones. For instance, in the more sparsely populated districts of Kericho and Nandi, farmers planted one maize crop per year and used late-maturing maize varieties. On the other hand, farmers in the densely populated areas of the same agroclimatic zone, such as Kisii and Nyamira Districts, attempted to grow two maize crops every year and planted a wider range of materials, including short- and medium-duration maize varieties (Hassan *et al.* 1992). For this reason, the population factor was taken into account in designing the maize survey.

The overlay of the agroclimatic zones and population density maps produced 32 climate–population strata to be surveyed. The area sown to maize in each stratum was obtained by overlaying the maize intensity of cultivation coverage with the climate–population classification of Table 3.1. The share of total maize area of each climate–population stratum was weighted by population density. Weights (w_i) were assigned to the four population density classes (i = 1,..., 4) such that a higher weight was assigned to a higher population density. An exception was the very high population density group, which was assigned the smallest weight because it represented urban centers where little or no maize was grown. A population density index (ϕ_i) was thus derived:

$$\phi_i = w_i/\Sigma_i w_i. \tag{3.1}$$

The maize area in each zone was indexed by the population density factor ϕ_i and a weighted percentage share of total maize area calculated (Table 3.2).

Several weighting schemes were applied to population density classes to reflect the importance of the number of people to be affected by targeted technological change. However, changing the weighting system caused no significant variation in the resulting weighted percentage area, mainly because of the direct relationship between population density and area sown to maize. With the exception of urban centers, more maize is grown where population

Table 3.2. Weighted percentage maize area by climate–population stratum, Kenya.

	Population density stratum[a]				
	Low	Medium	High	Very high	Total
Weight (w_i)[b]	0.50	0.75	1.00	0.25	2.5
Index (ϕ_i)[c]	0.20	0.30	0.40	0.10	1.0
Weighted percentage maize area[d]					
Agroclimatic zone					
Dry lowland tropics	0.70	0.25	0.07	0.00	1.02
Moist lowland tropics	1.44	0.55	0.22	0.01	2.23
Dry midaltitude	4.10	4.58	0.70	0.07	9.44
Moist midaltitude	6.11	27.07	11.94	0.8	45.91
Dry high-altitude	1.60	6.77	1.25	0.11	9.72
Moist high-altitude	5.72	14.99	6.25	0.28	26.24
Cool high-altitude	1.78	1.87	0.10	0.01	3.83
Extreme water stress	0.90	0.63	0.28	na	1.81
Total	21.22	56.71	20.80	1.28	100.00

[a] Low density = <100 persons km^{-2}; medium = 100–400 persons km^{-2}; high = 400–800 persons km^{-2}; and very high = >800 persons km^{-2}.
[b] Weights for density of population.
[c] Weight divided by sum of weights (2.5).
[d] Calculated as: $[\phi_i \times$ maize area in stratum $i]/[\Sigma_i\phi_i \times$ (maize area in stratum i)].

density is greater, indicating that demand for food is an important determinant of the cropping system and crop mix.

The sample size and allocation of sampling fractions

The statistical theory of sampling provides objective criteria for choosing an optimal sample size. As only part of a population is selected in a sample, surveys should be designed to achieve the highest possible efficiency in approximating attributes of the underlying population using the sample data. Sampling precision can be increased by either reducing variability between sampling units or by increasing the size of the sample. When resources are limited, researchers must weigh the gains in efficiency from larger samples against the costs associated with sampling extra units. In the absence of data on the degree of variability within the population of maize farmers for the relevant variables, an optimal sample size could not be determined for this survey. Instead, the sample size of 1200 farmers was determined on the basis of budgetary resources and time available for the survey.[2] However, to minimize variance, maize farmers were grouped into homogenous strata using climate and population density variables (Table 3.1).

After determining the sample size, it is critical to decide how many survey sites and farmers per site will allow an optimal allocation of the sample. As the main purpose of this survey was to obtain a detailed description of the spatial diversity in maize production practices and farming circumstances, capturing variability between sites was considered more important than capturing variability within sites. This priority implied surveying more sites and a smaller number of farmers per site. At the same time, the relative importance of site-specific conditions in defining target farmer domains and locations was taken into account through the spatial stratification by agroclimate, population density, and maize area. There are, however, many other socioeconomic factors and system parameters causing considerable variation within agroclimatic zones. If more sites are selected that are well distributed over the space, little variation should remain within sites, so a large number of respondents per site is not needed. Although the survey becomes more costly as the number of survey sites increases, it is important to note that as the number of sites diminishes, the variance between sites increases and the sample estimates become less efficient. In making an optimal choice between the number of sites and number of respondents per site, it is necessary to compare the relative cost and efficiency gains (smaller variance) of surveying more sites versus surveying more farmers per site – that is, determining which is cheaper and which contributes to smaller standard errors of sample estimates.

A sample of 20 farmers per survey site was considered adequate to capture within-site variability, which meant that 60 sites would be surveyed altogether (1200/20). Although the MDBP wished to increase the number of sites, researchers nevertheless felt that considerable variation remained between farmers and that it was necessary to sample the minimum of 20 farmers for each site. It was also considered statistically undesirable to have fewer than 20 data points (degrees of freedom) for analyzing variation between farmers, given the large number of farmer-specific variables included in the study. Moreover, should the survey design prove inadequate, it would be much easier and less expensive to incorporate more sites into the survey than to add more farmers per site. In other words, it would be more costly to correct for selecting fewer than the optimal number of farmers per site than for selecting a less-than-optimal number of sites. A post-survey analysis could also be done for various subsamples to test hypotheses about the optimal number of sites for the survey (see the discussion later in this chapter).

The 60 survey sites were allocated among the various climate–population strata in proportion to the weighted percentage share of maize area derived in Table 3.2. This allocation scheme allowed more farmers to be sampled from regions with higher concentrations of maize production and population. For a climate–population stratum to be allocated one survey site, its weighted percentage share of maize area had to exceed 1.7% (20/1200), the equivalent of 17,630 ha of maize.

A two-step procedure was used to allocate survey sites among the various climate–population strata. First, survey sites were allocated among the agroclimatic zones. Although the dry lowland tropics had less than a 1.7% weighted share of maize area and did not qualify for inclusion in the survey, this zone was purposely retained in the sample as a distinct maize production

Table 3.3. Allocation of sampling units among the selected climate–population strata, by weighted share of maize area, Kenya.

Agroclimatic zone	Sampling units by population density stratum[a]				Sampling units per agroclimatic zone	Number of farmers per agroclimatic zone
	Low	Medium	High	Very high		
Dry lowland tropics	1	1	0	0	2	40
Moist lowland tropics	3	1	0	0	4	80
Dry midaltitude	2	3	0	0	5	100
Moist midaltitude	4	17	7	0	28	560
Dry high-altitude	1	4	1	0	6	120
Moist high-altitude	3	9	4	0	16	320
Cool high-altitude	1	1	0	0	2	40
Extreme water stress	2	0	0	0	2	40
Total	17	36	12	0	65	1300

[a] Low density = <100 persons km^{-2}; medium = 100–400 persons km^{-2}; high = 400–800 persons km^{-2}; and very high = >800 persons km^{-2}.

environment in Kenya's Coast Region. It was decided to select a minimum of two survey sites from each agroclimatic zone entering the sample to capture possible spatial variations not controlled for by the climatic classification. Second, the survey sites were allocated among population classes within zones in proportion to their weighted percentage shares of maize area (Table 3.2).

Table 3.3 presents the allocation of survey sites among the climate–population classes. The number of sites to be selected from each stratum (Table 3.3) was obtained by dividing the weighted percentage share of maize area (Table 3.2) by the 1.7% required area per site. The exceptions were the dry lowland tropics and very dry zones (extreme water stress), where the minimum number of two sites per qualifying zone was applied. Moreover, in the moist lowland tropics the number of survey sites was increased to four to represent the three distinct maize production systems observed during informal surveys in the Coast Region (i.e., the cashew-, coconut-, and open maize-producing zones) and to allow for one site outside the coastal strip (Plate 1). Variation between the maize–cashew and maize–coconut systems resulted from differences in soil characteristics (Hassan *et al.* 1992). These changes raised the total sample size to 1300 farmers and 65 sites. None of the climate–population strata in the very high population density class qualified as sampling sites, for they mainly represented urban centers having a very small share of maize area.

Selection of respondents and survey sites

A multistage stratified random sampling procedure was adopted to select sites and farmers within each of the final climate–population strata chosen on the

Table 3.4. The National Sample Surveys and Evaluation Program (NASSEP III) clusters, Kenya, 1990.

Population group (number of people)	Number of districts and urban centers	Number of clusters per district[a]	Total number of clusters
Rural strata			
0–99,999	5	12	60
100,000–249,999	4	16	64
250,000–499,999	16	24	384
> 500,000	15	36	540
Urban strata			
All district headquarters and all towns with population of > 10,000 persons	58	–	324
Total	–	–	1372

Source: Compiled from Akach (1990).

[a] The total number of clusters per district varies in proportion to the density of population in the district (rural strata).

basis of the minimum weighted percentage share of maize area. The National Sampling Frame (NSF) developed by the National Sample Surveys and Evaluation Program III (NASSEP III) was used in the selection. The NASSEP III has 1048 rural and 324 urban clusters (Table 3.4) spread throughout about 40% of the sublocations in Kenya. For several reasons, the MDBP elected to adopt the NASSEP III clusters to sample maize farmers. First, NASSEP III clusters were widely distributed throughout the maize production zones. Second, the NSF provided the most up-to-date listing of Enumeration Areas, as well as names and occupations of household heads, sizes of holdings, and other information about households within each cluster.[3] Each cluster comprised more than 100 households. Third, NASSEP III data were available in digital form and could combine readily with the other data sets used by the MDBP. Finally, and most important, without the NASSEP III clusters, the MDBP would have spent considerable time and resources developing a country-wide sampling frame for selecting survey sites and maize farmers.

To locate survey sites within the selected climate–population strata, a random spatial search was applied on NASSEP III clusters within the strata. Because NASSEP III clusters are coded in monotonically increasing order across the various districts of Kenya, clusters could be ordered spatially using cluster codes. To distribute the selected clusters evenly across districts, systematic random sampling was used to select from the ordered list of cluster codes. A cluster was selected for every survey site.[4]

The households in the selected NASSEP III clusters were arranged in order of holding size after nonfarmers were dropped. A sample of 20 farmers and a

Table 3.5. Distribution of survey sites, by district, Kenya.

District	Number of sites	District	Number of sites
Kisii	3	Kakamega	2
Nyamira	2	Nandi	3(1)
Kericho	3(1)	E. Marakwet	1
Kisumu	2	Busia	2
S. Nyanza	3	TransNzoia	2(2)
Siaya	3	Uasin Gishu	2(2)
Nakuru	2(2)	Bungoma	2(1)
Narok	1(1)	West Pokot	2
Embu	2	Baringo	2
Nyeri	2	Kiambu	2
Kirinyaga	2	Nyandarua	2
Muranga	2	Laikipia	2
Meru	2	Makueni	2
Machakos	2	Kitui	3
Kilifi	3	Kwale	2

Note: Numbers in parentheses represent extra sites purposively selected from areas where large-scale commercial maize farming is done. Total number of districts covered = 30; total sites = 75; total number of farmers surveyed = 1400.

reserve of five farmers to provide for nonresponse were drawn from each cluster using systematic random sampling. The distribution of the selected survey sites (NASSEP III clusters) is presented in Map 2 and Table 3.5.

The survey of large-scale farmers

At survey time, it was observed that NASSEP III clusters included only small- to medium-scale farmers, who farmed areas of less than 100 acres (about 40 ha). Large commercial farms were probably excluded because of the design of the Central Bureau of Statistics (CBS) national sampling frame, not because of the random selection of clusters. The household rather than the farm is the sampling unit for the CBS population census, and large farm estates whose owners usually do not reside on the farm were not in the CBS clusters. Some clusters were adjacent to large estates, but they represented housing camps for the permanent laborers on the estates (Hassan *et al.* 1992).

To compensate for this anomaly, ten more clusters, each containing ten large-scale farmers, were purposively sampled in Kenya's large commercial farming areas, which increased the number of farms surveyed to 1400. Large-scale farmers' sites were located for geo-referencing using a Global Positioning System (GPS) device with a position fix accuracy of 25 m. More sites and fewer farmers per site were selected because less variation was expected between large-scale farms within sites than between sites.

According to the CBS crop focus surveys, about 25% of Kenya's maize is produced by large-scale farmers, who cultivate about 15% of the maize area (CBS 1990a). The CBS defines 'large-scale' as more than 20 acres (about 8 ha) of land. For its surveys, the MDBP assumed that half of the maize area under large-scale maize production (i.e., 7.5% of total area) is planted by farmers having 20–100 acres, and the other half (7.5% of total area) is sown by farmers having more than 100 acres. This assumption was the basis for including 100 farmers whose farms exceeded 100 acres (7% of total sample) in the survey sample. Large-scale commercial maize production occurs mainly in Nakuru, Uasin Gishu, and Trans Nzoia Districts, plus some parts of Narok, Kericho, Nandi, and Bungoma Districts, and these are the regions included in the survey of large-scale farmers (Table 3.5).

Although large-scale farmers cultivate 15% of the maize area, they constitute less than 15% of the maize farmers in Kenya. Estimates based on the proportion of farmers following a specific practice in this sample represent the percentage of maize area dedicated to that practice. Data from large-scale farmers thus need to be weighted to represent less than 15% of the population of maize farmers. If we do not control for this factor, parameter estimates of attributes such as average yield, amount of fertilizer applied, farm size, and percentage of farmers owning a tractor will be biased upward, towards the tendencies of large-scale farmers. A weight of 0.33 was used on data from large-scale farmers when percentages of farmers or estimates of quantitative variables were derived, based on the assumption that large-scale farmers constitute 5% (one-third of the 15% share in the sample) of the population of the maize farmers in Kenya.

The Trade-off between Numbers of Sites and Numbers of Farmers per Site

An attempt was made to evaluate the efficiency of sampling more sites versus more farmers per site. Ideally, prior knowledge about parameters characterizing the underlying distribution (mean and variability) of key attributes of the population under study enables sampling efficiency to be evaluated properly. As noted earlier, such information is usually lacking when surveys are designed. After the MDBP survey was done, the survey data were analyzed to compare parameter estimates that would have been obtained by using alternative survey designs. Subsampling from the selected cases and survey sites was employed to examine the efficiency trade-off between including more units versus more sites in the survey. Estimates of the following attributes of maize farming in Kenya were compared:

- Adoption rates of improved seed and fertilizer technologies, measured as the percentage of farmers using hybrid seed and amount of basal fertilizer applied to maize.

- Farmers' estimate of maize yields in a good season.
- Farmers' estimate of the average days to maize maturity.
- Intensity of maize cultivation (double-cropping).
- The cropping pattern (percentage of farmers planting maize with other crops).
- The gender of the head of the farm household.
- The percentage of farmers using animal manure.
- Contact with extension.

Given the objectives of this survey, these attributes were considered the most important factors characterizing maize production in Kenya and regarded as key determinants of the potential role of maize technology development and transfer in improving productivity. Various combinations of fewer sites and numbers of farmers per site were explored. Random sampling was used to select 15 and 10 of the 20 farmers surveyed at each site (representing a 75% and 50% reduction in the sample size). Subsamples of 75% and 50% of the selected sites were randomly drawn from the original sample of 75 sites.

Similar results were obtained from simple versus systematic random sampling of farmers and survey sites. Simple statistical procedures were used to test hypotheses about significant deviations from original sample estimates (Table 3.6). Results indicate that estimates obtained from 15 farmers per site (75% reduction) or from 75% of the sites (with 20 farmers each) were not significantly different from original estimates based on 100% of the selected cases. Deviations from original sample estimates became significant as the number of farmers or sites was reduced by 50%. It is also clear that deviations were larger when fewer sites were selected than when fewer farmers were considered. However, the cost of interviewing an extra farmer at a site is much lower than the cost of surveying one more site.

These results suggest that if the attributes analyzed in Table 3.6 represent satisfactory criteria for making decisions about sampling densities, very little would have been sacrificed in terms of sampling efficiency by interviewing between 10 and 15 farmers in each of the 75 sites surveyed. At the same time, the results indicate that efficiency gains from surveying more sites would have been insignificant. It is important to note, however, that all of the above depends on the adequacy of the stratification scheme adopted to represent distinct maize production conditions and groups of farmers. According to Table 3.6, it appears that a sampling fraction of 15 farmers per site, from 75% of the sites included in the original survey, would have been a more cost-effective design for studying the listed attributes. However, variable sampling fractions allocated between sites in proportion to the percentage of land under maize, population, or variability could be more appropriate than the uniform sampling fraction applied to all sites in the present design, in which an equal number of farmers is sampled from all sites. This is because some sites may be more homogeneous than others with respect to the attributes under study. We did make use of variable sampling fractions in our survey, however, to allocate total sample size between agroclimatic zone and population strata in proportion to the percentage area under maize.

Table 3.6. Effect of lower sampling densities within sites versus fewer survey sites for the national survey of maize farmers, Kenya.

Attribute	Total sample (100%)[a]	Within-site densities[b] 75%	Within-site densities[b] 50%	Number of survey sites[c] 75%	Number of survey sites[c] 50%	75% of sites[d] 75% density	75% of sites[d] 50% density
Yield							
Mean (t ha^{-1})	2.57	2.52	2.40	2.67	2.84	2.39	2.32
Standard deviation	2.26	2.21	2.05	2.25	2.13	1.99	1.91
t-value	–	0.50	1.61*	1.11	2.73**	2.00**	2.27**
Basal fertilizer							
Mean (kg ha^{-1})	53	55	64	60	62	62	70
Standard deviation	160	179	196	175	145	202	238
t-value	–	0.30	1.24*	1.02	1.32*	1.06	1.49*
Time to maturity							
Mean (days)	174	175	177	171	168	172	164
Standard deviation	57	57	55	56	59	56	57
t-value	–	0.4	1.12	1.29*	5.32*	0.78	3.37*
	Percentage						
Adoption of hybrid maize seed (% farmers using)	69	68	74	68	76	65	62
Double cropping (% farmers)	41	41	44	51	58	46	56
Intercropping (% farmers)	78	77	74	75	71	72	74
Gender of household head (% female)	41	44	40	41	37	45	46
Use of manure (% farmers)	55	55	59	55	48	45	41
Never received extension advice (% farmers)	40	39	36	43	46	40	38
Number of cases	1305	979	653	1097	768	770	513

Note: * indicates significant at 5%; ** at 1%.

[a] The total number of cases and sites in the original sample, excluding the surveys of large-scale farmers (102 farmers).

[b] Denotes the percentage reduction in number of cases interviewed per site – i.e., 15 (75%) and 10 (50%).

[c] Percentage reduction in number of sites surveyed.

[d] Fifteen and ten farmers were selected randomly in each of the 75% of the sites in the original sample.

Conclusion

The approach described here for designing a geo-referenced survey of maize farmers in Kenya was intended to enhance researchers' ability to target distinct production situations and technology adaptation needs. The resulting survey data can be integrated with experimental and other data for direct manipulation using GIS. The data sets compiled and methods developed in the course of this work are also easy to modify for other purposes, especially for planning commodity based research and building spatially integrated information systems. The next chapter offers an example of how information from the survey of Kenyan maize farmers was used to analyze the suitability of the agroclimatic classification of maize production zones used to structure maize research in Kenya.

References

Akach, S.O. (1990) The National Sample Surveys Program (NASSEP). In: *Proceedings of the Symposium of Producers and Users of Statistics.* UNDP/UNDTCD and Central Bureau of Statistics, Ministry of Planning and National Development, Nairobi, Kenya, pp. 54–68.

Carter, S., and Jones, P. (1989) *COSCA Site Selection Procedure.* Collaborative Study on Cassava in Africa Working Paper No. 2. International Institute of Tropical Agriculture (IITA), Ibadan, Nigeria.

CBS (Central Bureau of Statistics). (1990a) *Crops Focus Survey.* Ministry of Planning and National Development, Nairobi, Kenya.

CBS (Central Bureau of Statistics). (1990b) *Pre-Population Census Coverage of Households by Sub-location.* Ministry of Planning and National Development, Nairobi, Kenya.

CBS (Central Bureau of Statistics). (1991) *Statistical Abstract.* Ministry of Planning and National Development, Nairobi, Kenya.

Hassan, R., and Karanja, D. (1992) *Development of Survey Questionnaires and Organization of the Field Work.* KARI/CIMMYT Maize Data Base Project Document No. 5. Kenya Agricultural Research Institute (KARI) and International Maize and Wheat Improvement Center (CIMMYT), Nairobi, Kenya.

Hassan, R.M., Corbett, J., Karanja, D., and Okoth, P. (1992) *The Spatial Sampling Frame and Design of Farmers' Survey.* KARI/CIMMYT Maize Data Base Project Document No. 4. Kenya Agricultural Research Institute (KARI) and International Maize and Wheat Improvement Center (CIMMYT), Nairobi, Kenya.

Lazerwitz, B. 1971. *Sampling Theory and Procedures.* McGraw-Hill, London, U.K.

Otichillo, W.K., and Sinange, R.K. (1991) *Long Rains Maize and Wheat Production in Kenya in 1990.* Technical Report No. 140. Department of Resource Surveys and Remote Sensing, Ministry of Planning and National Development, Nairobi, Kenya.

Pollak, L.M., and Pham, H.N. (1989) Classification of maize testing locations in sub-Saharan Africa by using agro-climate data. *Maydica* 34, 41–43.

Endnotes

1 The total land area of Kenya was 57 million hectares in 1990 (CBS 1991), which means that about 2% of it was sown to maize during the long rains of 1990 (i.e., 1074/57,000) (Table 3.1).

2 When the farmer survey was planned, the cost of surveying one farmer was estimated at US$ 25, which, apart from reflecting the cost of conducting the survey, included planning expenses such as conducting informal surveys, training enumerators, and pretesting questionnaires. One interview was estimated to take an hour and a half to complete. Given an estimated average traveling distance of 150 km to and from survey sites, it was estimated that one enumerator could interview only two farmers per day (Hassan and Karanja 1992).

3 Updated for 1990–1994 using the 1989 population census data (Akach 1990).

4 In the absence of a well-developed, up-to-date, and consistent sampling frame such as NASSEP III, the MDBP would have to have taken another approach in selecting survey sites. For example, the Collaborative Study of Cassava in Africa (COSCA) used a random grid-search approach to select survey sites. In this procedure, the survey area (cassava-growing countries in Africa) was divided into equal-sized grid cells which were numbered consecutively from one to *N* (Carter and Jones 1989). Systematic random sampling was applied to select a grid cell for every survey site. Whereas the grid-search approach is free from various subjective sampling biases and inadequacies that affect many of the available sampling frames, it involves extra sampling stages, such as the selection of villages within grid cells. The wide geographic coverage requires the demarcation of large grid cells that contain many settlements/villages – hence the extra step of developing a list of villages to select from for each of the selected grid cells. Other problems associated with the grid-search approach (e.g., there may be no settlement within a selected grid cell; some sites may be inaccessible) are discussed in Carter and Jones (1989). However, it is not easy to find a consistent sampling frame such as that offered by NASSEP III for an area as large as that covered by the COSCA survey.

4 Combining Geo-referenced Survey Data with Agroclimatic Attributes to Characterize Maize Production Systems in Kenya

RASHID M. HASSAN, JOHN D. CORBETT, AND KIARIE NJOROGE

Data from the nationwide survey of maize farmers were combined with other spatial data to refine the initial classification of maize adaptation zones and characterize maize production systems in Kenya. Analyses of the determinants of farmers' planting strategies confirmed that farmers' choice of varieties deviated from varieties recommended by researchers for the maize adaptation zones used by the national maize program. One interpretation of this discrepancy is that farmers grow other varieties because the time to maturity of the recommended varieties is actually not suitable to farmers' conditions. The divergence was highest in the midaltitude zone. New boundary conditions were developed to distinguish a transitional zone between the typical midaltitude and highland tropical zones. Areas encompassed by the new zone were slightly cooler than areas belonging to the midaltitude zone and warmer than those in the highland tropics. The new transitional zone and the midaltitude zone encompass most of Kenya's maize area and maize farmers, and should receive greater attention from the maize research system. The major characteristics of Kenya's maize farming systems include the dominance of intercropping; cropping intensity; various biotic and abiotic stress factors; and the influence of population density on intensity of maize cultivation, time to maturity of germplasm used by farmers, and pressure on fragile lands.

Crop production environments are classified into agroclimatic zones to separate spatial domains having different needs for technology and responding differently to technological interventions. This kind of classification is crucial for defining the limits of potential gains from biological research and for planning and targeting agricultural innovations effectively. The maize adaptation zones delineated by the MDBP (Chapter 2) were developed on the basis of climatic

 Maize Technology Development and Transfer (ed. R.M. Hassan)

suitability and experts' definition of adaptability ranges for maize germplasm. However, biophysical suitability alone does not guarantee that a technology will be successful. Numerous farmer-specific, socioeconomic, institutional, and policy factors interact with the physical environment and influence the transfer of technology. For this reason, the MDBP sought to combine information collected from farmer surveys (Chapter 3) with other spatial data to refine and consolidate the initial classification of maize adaptation zones and to characterize maize production systems in Kenya. The ultimate objective of this work is to be able to design appropriate technologies that are better targeted to meet farmers' needs and thus adopted more rapidly.

The Present Classification of Maize Production Environments

The national maize research program in Kenya was organized to serve five major production environments: coastal areas (lowland tropics), moisture-stressed midaltitude areas, midaltitude areas without moisture stress, high-altitude areas, and very-high-altitude areas (Gebrekidan *et al.* 1992). This classification is consistent with the International Maize and Wheat Improvement Center's (CIMMYT's) classification of maize production environments in developing countries. CIMMYT separates production environments into 'mega-environments,' which are large, not necessarily contiguous areas having similar requirements for maize production (e.g., areas where the maize crop is exposed to similar biotic and abiotic stresses). CIMMYT's three primary maize mega-environments are based on altitude zones (CIMMYT 1988): the low-altitude (lowland tropical) mega-environment, midaltitude (subtropical) mega-environment, and high-altitude (highland tropical) mega-environment. While such a classification may be suitable for an international center having a global mandate for maize research, it should be further subdivided to be useful at national and regional levels. As noted earlier, KARI refined this mega-environment scheme using rainfall as a criterion to subdivide the midaltitude zone. When the MDBP developed its initial classification of maize production environments, the dry–moist regimes were extended to other mega-environments and a new zone characterized by extreme water stress was added to the classification (Geographical Data Table 1 and Plate 1).[1] The largest area sown to maize in Kenya in 1990 was in the midaltitude zone, followed by the highland tropics (Geographical Data Table 1).

Data from the farmer survey, which extended across 30 districts, were used to examine whether farmers in each maize production zone delineated by the MDBP actually used maize materials whose characteristics matched the characteristics considered appropriate for the zone. The adequacy of the germplasm maturity requirements for the MDBP zones was evaluated against data on adoption of recommended maize varieties, farmers' planting strategies, and the length of the growing season.

Adoption of recommended genotypes

Improved maize varieties released by KARI were grouped into four categories:

- Late-maturing materials (600-series hybrids).
- Medium-maturing materials (500-series hybrids).
- Dryland materials (short time to maturity; drought tolerant).
- Lowland tropical (coastal) materials.

Maize researchers had targeted these different types of germplasm to the different adaptation zones, as shown in Table 4.1. An analysis of data from the farmer survey revealed that in all zones except the highland tropics, farmers' selection of maize varieties did not correspond to the classification in Table 4.1. These inconsistencies were particularly noticeable in the midaltitude zone, where a large percentage of farmers used varieties developed for other environments. For example, more than 70% of the maize area in the midaltitude zone was sown to late-maturing materials (600-series hybrids), whereas the medium-maturing (500-series) hybrids targeted for that environment occupied only 12% of the land under maize (Table 4.1). While 84% of farmers in the lowland tropics used local materials rather than the recommended improved varieties, most farmers in the midaltitude and highland zones purchased improved seed (61% and 93%, respectively). The high use of local varieties in the lowland tropics could be attributed to farmers' lack of awareness of improved varieties, lack of access to them, or the cost of improved maize seed. However, none of these factors explains farmers' preference for late-maturing (600-series) hybrids over the medium-maturing (500-series) hybrids in the midaltitude zone, because both kinds of seed are available and sell for the same price.

Other explanations for the inconsistency between the agroclimatic classification of Table 4.1 and farmers' behavior could also be that the agroclimatic classification inadequately characterized maize germplasm and maturity requirements or that farmers are not satisfied with the varieties recommended for their production environment.

Improving the present classification

Clearly, the classification of maize production environments needed to be refined to reflect farmers' maize production practices more accurately. New boundary conditions were developed on the basis of the perception that the transition between the lowland tropical and midaltitude environments seemed to be captured adequately, whereas the boundary between the midaltitude and highland tropical environments was more difficult to establish. Because the transition from a typical midaltitude to a typical highland tropical environment extends over a large area in which climate attributes change gradually, it is difficult to distinguish an abrupt boundary between the two zones. Regions falling within this transitional range were cooler than the 'typical' midaltitude environment but warmer than 'typical' highland tropical environments. Since temperature rather than elevation *per se* determines plant growth

Table 4.1. Distribution of maize area, germplasm, and sample farmers in Kenya, by agroclimatic zone.

Agroclimatic zone	Area under maize hybrids: 600 series (%)[a]		Area under maize hybrids: 500 series (%)[a]		Area under improved dryland materials (%)[b]		Area under improved coastal materials (%)[c]		Area under unimproved materials (%)[d]		1990 maize area (000 ha)[e]	Percentage of total maize area	Percentage of farmers in sample
Lowland tropics	0	(0)	0	(0)	6	(6)	10	(10)	84	(84)	42	4.1	7.1
Dry lowland tropics	0	(0)	0	(0)	5	(8)	1	(3)	94	(89)	13	1.3	2.8
Moist lowland tropics	0	(0)	0	(0)	6	(5)	23	(17)	71	(78)	29	2.8	4.3
Midaltitude zone	76	(49)	12	(10)	1	(2)	0	(0)	21	(39)	566	54.6	53.8
Dry midaltitude zone	7	(12)	6	(10)	22	(12)	0	(0)	65	(66)	110	10.6	10.0
Moist midaltitude zone	82	(58)	2	(10)	**	(1)	**	(1)	16	(30)	456	44.0	43.8
Highland tropics	84	(87)	13	(6)	0	(0)	0	(0)	5	(7)	408	39.3	33.5
Dry highland tropics	79	(69)	15	(21)	0	(0)	0	(0)	6	(10)	100	9.6	7.5
Moist highland tropics	88	(93)	12	(3)	0	(0)	**	(1)	3	(8)	265	25.6	21.6
Cool highland tropics	94	(89)	**	(2)	0	(0)	0	(0)	6	(9)	43	4.1	4.4
Extreme water stress zone	0	(0)	**	(1)	26	(31)	0	(0)	74	(69)	21	2.0	5.6
Total	61	(55)	9	(7)	2	(4)	1	(1)	27	(33)	1037	100.0	100.0

Note: numbers in parentheses indicate the percentage of farmers using the cultivars; ** indicates a maize area of less than 0.5%.
[a] The 600-series hybrids have a longer time to maturity and the 500-series hybrids have an intermediate time to maturity.
[b] Improved dryland varieties are Katumani and Makueni Composites.
[c] Improved coastal materials are Coast Composite and Pwani Hybrid. Coastal materials were developed particularly for lowland tropical environments.
[d] Total area may sum to more than 100 because some farmers use more than one variety.
[e] From Otichillo and Sinange (1991).

and drives development, this shift in temperature has important implications for the time required for maize to mature, the expected spectrum of diseases and pests in the environment, and, ultimately, the suitability of different types of germplasm for the environment.

Accordingly, a transitional zone between the midaltitude and highland tropical environments was added to the classification. More than 80% of the area in this zone was drawn from midaltitude areas and the rest from highland tropical areas. Elevation, temperature, and precipitation criteria used to describe and cluster the new set of maize-specific agroclimates are provided in Geographical Data Table 2 and their spatial distribution shown in Plate 3. From Geographical Data Table 2 it may also be seen that each of the new zones is adequately represented in the sample of farmers.

The percentage of farmers using varieties targeted for the dry parts of the midaltitude zone increased significantly under the new classification (Table 4.2). Most farmers (78%) in the moist portion of the transitional zone used late-maturing materials. Their practices were similar to those of typical farmers in the highland tropics. However, the percentage of farmers attempting two maize crops per year is higher in the transitional zone than in the highland tropics. In the drier parts of the transitional zone, which have the highest percentage of farmers (76%) growing two maize crops every year, farmers prefer medium-maturing materials (500-series hybrids) to the 600-series hybrids favored in the highland tropics (Table 4.2). It is also important to note that under the new zonal classification the percentage area planted to unimproved varieties in the typical midaltitude zone doubled, rising from 21% to 42% (Tables 4.1 and 4.2). At the same time, most of the land (82%) in the transitional zone was sown to improved maize varieties. This confirms once more that the transitional and typical midaltitude zones represent distinct maize production environments.

Nevertheless, even under the new zonal classification, farmers in the midaltitude zone tended to use late-maturing hybrids, indicating some degree of dissatisfaction with the medium-maturing (500-series) cultivars (Table 4.2). A clear pattern was also observed in the use of the medium- and late-maturing materials within and between the midaltitude, transitional, and highland tropical zones. A combination of early- and medium-maturing varieties dominated the drier part of the midaltitude zone, whereas the late-maturing varieties dominated the moist transitional and highland tropical zones. The varietal hybrid H614 was the most commonly used cultivar, followed by conventional inbred-derived hybrids H625 and H626.[2] Although conventional hybrids were not used in the midaltitude zone, many farmers used the varietal hybrid H614. In fact, more farmers in this zone preferred H614 over H512. These results suggest that H614 is better suited to farmers' conditions and agronomic constraints than H625 and H626 (Table 4.2), possibly because of H614's wider genetic base. (For a detailed analysis of maize farmers' varietal preferences, see Chapter 7).

Farmers outside the midaltitude zone are also dissatisfied with the available maize varieties. In the lowland tropics and relatively drier areas (extremely water stressed and dry transitional zones), a significant proportion

Table 4.2. Maize germplasm development and adoption by agroclimatic zone, Kenya.

Agroclimatic zone	Area under maize hybrids: 600 series (%)[a]	Area under maize hybrids: 500 series (%)[a]	Area under improved dryland materials (%)[a]	Area under improved coastal materials (%)[a]	Area under unimproved materials (%)	Most common improved varieties[b]	Area double-cropped (%)[c]	Number of improved varieties released	Releases in past ten years	Year of most recent release
Lowland tropics	0 (0)	0 (0)	6 (6)	10 (10)	84 (84)	KCB, CCB	25 (35)	2	1	1989
Dry lowland tropics	0 (0)	0 (0)	5 (8)	1 (3)	94 (89)	–	8 (20)	–	–	–
Moist lowland tropics	0 (0)	0 (0)	6 (5)	17 (23)	71 (78)	–	36 (46)	–	–	–
Midaltitude zone	22 (25)	13 (11)	9 (6)	0 (0)	56 (58)	–	42 (57)	4	–	–
Dry midaltitude zone	7 (16)	6 (10)	22 (16)	0 (0)	65 (58)	KCB	32 (50)	2	1	1989
Moist midaltitude zone	32 (29)	18 (11)	** (1)	0 (0)	49 (59)	H511, H614	47 (60)	2	0	1970
Transitional zone	82 (67)	6 (9)	0 (0)	** (1)	16 (13)	–	16 (45)	NT	NT	NT
Dry transitional zone	2 (3)	10 (18)	0 (0)	0 (0)	88 (79)	H511	79 (76)	NT	NT	NT
Moist transitional zone	87 (78)	7 (7)	0 (0)	** (1)	6 (15)	H614	10 (40)	NT	NT	NT
Highland tropics	86 (82)	11 (6)	0 (0)	** (0)	5 (14)	H614	3 (22)	10	5	1989
Dry highland tropics	89 (78)	3 (10)	0 (0)	0 (0)	8 (12)	–	7 (26)	–	–	–
Moist highland tropics	87 (88)	16 (6)	0 (0)	0 (0)	7 (9)	–	3 (24)	–	–	–
Cool highland tropics	94 (87)	** (2)	0 (0)	0 (0)	6 (11)	–	1 (5)	–	–	–
Extreme water stress zone	0 (0)	** (1)	26 (31)	0 (0)	74 (69)	KCB	43 (44)	NT	NT	NT
Total	61 (55)	9 (7)	2 (4)	1 (1)	27 (33)	H614, H625	38 (41)	–	–	–

Note: numbers in parentheses indicate the percentage of farmers who have adopted the type of cultivar or practice; 'NT' indicates environments not targeted by the breeding program; and ** indicates a maize area of less than 0.5%. As some farmers mixed varieties, totals add to more than 100% in some cases.

[a] The 600-series hybrids have a relatively long time to maturity and the 500-series hybrids have an intermediate time to maturity. Dryland materials are early maturing. Coastal materials were developed for lowland tropical environments.

[b] KCB = Katumani Composite; CCB = Coast Composite; H511, H614, and H625 are hybrids.

[c] Percentage area sown to maize in both seasons of the year; the percentage farmers who double-crop is indicated in parentheses.

of farmers used local varieties (Table 4.2). The reasons for farmers' dissatisfaction may lie with the historical pattern of varietal development within KARI and the total number of maize cultivars released. Table 4.2 clearly shows KARI's historical bias in favor of highland tropical environments. Not a single new variety of medium maturity has been released for the moist midaltitude zone over the past 20 years. Over the past ten years, only one such variety was released for the lowland tropics and one for the dry midaltitude zone, compared with five new releases for the highland tropics. No variety specifically targeted for extremely water-stressed conditions is yet available, and many farmers in the midaltitude and transitional zones plant a combination of medium- and late-maturing cultivars, perhaps because a suitable variety is lacking or perhaps as a way of managing risk. Nevertheless, these findings suggest that KARI should shift its maize research strategy and resources towards these relatively neglected environments, where most of the maize area and farmers are found.

Maize Maturity in the Agroclimatic Zones and Farmers' Planting Practices

Farmers' planting practices are as important as germplasm suitability and use for evaluating whether the agroclimatic zones adequately reflect maize production systems. The timing of maize planting and harvest and the intensity of cropping are determined by the prevailing socioeconomic and physical environments and strongly influence farmers' choice of variety. Farmers often attempt two maize crops per year in areas where rains are bimodal and population density high. On the other hand, in areas where land is relatively abundant (favoring extensive cultivation) and there is a long single rainy season, farmers opt for a single crop of late-maturing maize.

The influence of various biophysical and socioeconomic factors on farmers' planting strategies and the maturity of maize varieties used by farmers can be examined for each agroclimatic zone, and key attributes affecting variation between and within zones can be identified to determine whether the zones should be refined further. For this analysis, the ten agroclimatic zones described in Geographical Data Table 2 (see also Plate 3) were grouped into six main zones (shown in Plate 4). As no significant variation was observed between the extremely water-stressed and dry midaltitude zones (Table 4.2), they were combined into a dry midaltitude zone. Given the relative similarity in farmers' practices within the highland tropics, the dry, moist, and cool segments of this zone were combined to form a single highland tropical zone. The same was done for the lowland tropical zone. However, the moist midaltitude and dry and moist transitional zones were maintained as separate environments, mainly because farmers' practices, production systems, and agroclimatic stress factors were different in each zone. Information on maize maturity and planting is summarized in Tables 4.3 and 4.4 for these six agroclimatic zones.

Dry midaltitude zone (early-maturing germplasm)

The dry midaltitude zone is characterized by low and highly variable rainfall. While half of the farmers in this zone used a local maize variety known as White Machakos Local, the improved variety, Katumani Composite, was also popular, especially among farmers who produced a single maize crop (Table 4.3). Given that Katumani Composite is an open-pollinated variety (OPV) released more than 25 years ago in this region, there is a very high chance that what farmers refer to as Machakos Local is actually an advanced generation of Katumani, often mixed with Machakos, rather than the totally unimproved local material.[3]

There are important differences between the dry midaltitude and dry transitional zones. Whereas low and erratic rains are common in both zones, they differ significantly in elevation and hence temperature (Geographical Data Table 2), density of human population, reliability of the March rains, and the percentage of farmers who double-crop maize (Table 4.3). These factors strongly influence farmers' planting strategies and have important implications for the kinds of germplasm they will need. The same maize cultivars (e.g., hybrid H511 and Machakos Local) did not mature as rapidly in the dry transitional zone as in the semiarid conditions typical of the dry midaltitude zone (Table 4.3). Thus maize maturity requirements vary significantly between the two zones.

Moist midaltitude and dry transitional zones (medium-maturing germplasm)

Two of the six agroclimatic zones were classified as areas suitable for medium-maturing maize germplasm. These areas include locations in the Lake Victoria region, which constitute the moist midaltitude zone, and the higher parts of Machakos and lower Kiambu Districts, which constitute the dry transitional zone (Plate 4). These zones differ in some important ways, however. Maize hybrids of intermediate maturity (500 series) were used more in the dry transitional zone, especially by farmers growing two maize crops (Table 4.3). The lake region local varieties[4] Rachar and Nyamula were the most common cultivars in the moist midaltitude zone, and late-maturing maize hybrids (600 series) were preferred to the 500-series hybrids by farmers who grew single maize crops in this zone.

The dry transitional zone, higher in elevation and relatively cooler than the warmer moist midaltitude zone, is expected to favor maize materials with a relatively longer time to maturity. The temperature effect on maize development was evident for varieties of intermediate maturity (e.g., H511), which needed less time to mature in the moist midaltitude zone compared with the dry transitional zone. The reverse was true for unimproved materials. The lake region local cultivars planted by farmers in the moist midaltitude zone took many more days to mature than the upland locals planted in the dry transitional zone.

The fact that farmers in the moist midaltitude zone, especially those double-cropping maize, planted lake region local maize varieties that take

Table 4.3. Recommended maturity class of maize germplasm and farmers' practices in the major maize season, by agroclimate and cropping intensity (n = 1407), Kenya.

Agroclimatic zone and recommended germplasm maturity class	Recommended varieties[a]	Days to maturity (exptl.)[b]	Farmers double-cropping maize: Most used varieties (% farmers)[a]	Farmers double-cropping maize: Average days to maturity[c]	Farmers single-cropping maize: Most used varieties (% farmers)[a]	Farmers single-cropping maize: Average days to maturity[c]	Average time to maturity[c]: Days[c]	Average time to maturity[c]: Thermal time (°C days)[d]	Percentage farmers double-cropping maize	Average population density (pers. km^{-2})[e]	Percentage farmers regarding March rains as the major season
Lowland tropics Early–intermediate	HPON CCB	90–120 120–150	CLOC (84) HPON (7)	119 (19) 87 (37)	CLOC (70) KCB (13) CCB (10)	122 (36) 95 (27) 130 (34)	120 (33)	2124	35	121	99
Dry midaltitude zone Early	KCB, MCB	90–120	KCB (14) MLOC (55) H511 (11)	129 (28) 129 (29) 130 (41)	KCB (41) MLOC (48) H511 (10)	81 (36) 92 (41) 111 (42)	114 (47)	1767	50	210	48
Moist midaltitude zone Intermediate	H511, H512, H622	120–150	LLOC (57) H511 (16)	161 (25) 135 (28)	LLOC (35) H511 (12) H614 (26)	160 (19) 132 (43) 218 (51)	163 (40)	2461	60	310	96
Dry transitional zone Intermediate	H511, H512	120–150	MLOC (35) ULOC (32) H511 (29)	130 (32) 138 (14) 136 (32)	MLOC (33) ULOC (17) H511 (50)	148 (8) 148 (34) 144 (25)	144 (20)	1829	76	398	46
Moist transitional zone Intermediate–late	H632, H622 H614, H625	150–180 > 180	H614 (24) H625 (26) H511 (16)	177 (37) 172 (26) 137 (26)	H614 (52) H625 (27) H626 (15)	201 (36) 189 (30) 198 (55)	181 (39)	2063	40	331	98
Highland tropics Late	H614, H625, H626	> 180	H614 (37) H625 (44) H511 (13)	191 (57) 176 (34) 176 (51)	H614 (53) H625 (16) H626 (18)	231 (50) 205 (43) 213 (66)	213 (53)	2066	22	238	89

Note: the semiarid zone encompasses the dry midaltitude and extreme water stress zones; the moist midaltitude zone encompasses the lake region.

[a] Varieties abbreviated as follows: KCB = Katumani Composite; CCB = Coast Composite; MCB = Makueni Composite; CLOC = Coast Local; MLOC = Machakos Local; LLOC = Lake Region Local; ULOC = Upland Local; 'H' denotes hybrids (HPON = Pwani).

[b] The range in days to maturity reflects the effect of within-zone variation in altitude on maturity (KARI 1992).

[c] Based on time of planting and farmers' estimates of the date when the maize crop reached maturity (i.e., was ready to harvest). Figures in parentheses are the percentage coefficient of variation.

[d] The sum of daily mean temperatures between zero and maximal rate of maize development (see Appendix A for details).

[e] Average of population density in survey sites within the zone (i.e., sample average), according to the 1989 population census data.

Table 4.4. Recommended and farmers' maize planting strategies for the March–August season by agroclimatic zone, cropping intensity, and size of holding (*n*=1407), Kenya.

Agroclimatic zone and recommended planting date[a]	Date when 80 mm rainfall accumulates[b]	Variability in total seasonal precipitation (CV %)[c]	Farmers single-cropping maize			Farmers double-cropping maize			Percentage farmers double-cropping maize	
			Percentage farmers plant within recommended dates	Percentage farmers plant earlier	Percentage farmers dry plant	Percentage farmers plant within recommended dates	Percentage farmers plant earlier	Percentage farmers dry plant	Large (> 8 ha)	Small (< 8 ha)
Lowland tropics April (1) – May (1)	April (1)	36	47	53	9	62	38	26	0	36
Dry midaltitude zone March (2) – April (1)	March (3)	52	87	13	62	45	55	33	39	50
Moist midaltitude zone April (1) – May (1)	March (1)	32	4	96	4	5	95	6	0	61
Dry transitional zone March (2) – April (3)	March (3)	40	89	11	78	66	34	42	–	76
Moist transitional zone March (4) – April (4)	March (1)	27	20	80	22	5	95	18	2	47
Highland tropics March (4) – April (1)	March (2)	32	29	71	20	10	90	21	1	26

Note: figures in parentheses denote the week of the month (1 = first week, 2 = second week, and so on).

[a] Recommendation for optimal planting date based on experimental results (KARI 1992).

[b] Date calculated based on long-term rainfall data for Kenya.

[c] Percentage coefficient of variation for mean total precipitation during the March–August season (the 'long rains'), calculated based on long-term rainfall data for Kenya.

relatively longer to mature than the 500-series hybrids indicates that the local varieties were preferred for qualities other than time to maturity. Better tolerance to low soil fertility levels and biotic stresses (such as *Striga* spp., a parasitic weed) may be more important to maize farmers in this zone than earliness for double-cropping. Results from field experiments have shown that in this zone selection of a tolerant genotype was more important than date of planting in reducing yield losses in maize from *Striga* infestation (Ransom and Osoro 1991).

In the dry transitional zone, on the other hand, farmers' preference for materials that matured in a relatively shorter period can be attributed both to the relative importance of the second season in that area and to the general requirement for rapid maturity as a drought-evading technology. Whereas all farmers in the moist midaltitude zone considered the March rains to be their major maize-growing season, only 46% of farmers in the dry transitional zone indicated the same (Table 4.3). More than half of these farmers thought the October rains were more reliable. This perception may explain why, for the March rains, farmers plant varieties that mature relatively early: this strategy allows their major maize crop to be planted on time in October. This perception may also explain why so many farmers attempt to produce two maize crops in the dry transitional zone. Most farmers (78%) dry plant before the March rains (Table 4.4). The relatively higher population density in this zone (Table 4.3) is another reason for the high rate of double-cropping.

These results suggest that important biophysical and socioeconomic characteristics differentiate the moist midaltitude and dry transitional zones as distinct production environments and distinct target domains for maize technology in Kenya. The zones should not be considered one homogeneous environment where the response to technological and policy interventions will be similar.

Moist transitional and highland tropical zones (late-maturing germplasm)

Late-maturing maize germplasm is used by the majority of farmers in the moist transitional and highland tropical zones (Table 4.3). However, the frequency of double-cropping is much higher in the moist transitional zone, possibly because of relatively higher population pressure and the bimodal rainfall pattern (in the highland tropical zone, rainfall extends over a single long season between March and November). In addition, the warmer temperatures of the moist transitional zone allow for two maize cropping seasons. These climatic factors (bimodal rains and warmer temperatures) and the relatively higher pressure on the land are incentives for farmers to plant two maize crops per year, but farmers lack a suitable genotype. The 600-series hybrids yield better than the 500-series and are better suited to the wet conditions of the zone during the cropping season that begins with the March rains. However, the longer time to maturity of the 600-series hybrids makes it difficult for farmers to attempt a second crop. Moreover, silking in these hybrids usually coincides with the onset of the short rains in August, resulting in heavy losses from rotting (Njoroge *et al.* 1992a).

Even so, the later maturing 600-series hybrids are preferred to the 500-series hybrids, even by double-croppers in the moist transitional zone (Table 4.3). Medium-maturing hybrids do not fit the short rains of this zone very well and are usually hit by blight (Njoroge *et al.* 1992a). Despite the fact that farmers can choose among a number of varieties available for the moist transitional zone, a suitable variety (i.e., of optimal maturity between medium and late cultivars) remains to be developed for this environment.

The lowland tropics

Maize breeders regard the hot humid environment of the Kenyan coast as a distinct production zone. Cultivars adapted to this environment and its range of biotic and abiotic stresses differ from those needed in other maize production zones (Njoroge *et al.* 1992b). Most farmers in this zone use unimproved maize genotypes (Table 4.3) whose average maturity ranges under farmers' conditions are consistent with ranges of recommended varieties. The local coast varieties are slower to mature than the Coast and Katumani Composites, which are mainly used by farmers growing one maize crop per year. The varietal hybrid Pwani, recommended for the lowland tropical zone, is grown only by farmers attempting two maize crops per year. A separate study using the same survey data (Chapter 7) showed that the high use of unimproved maize germplasm in the lowland tropics results mainly from lack of access to other materials. No seed dealers were found in any of the villages surveyed in this zone.

Average time to maturity of maize cultivars in relation to farmers' planting practices

Average time to maturity (crop duration from planting to harvest) is expressed in calendar days as well as thermal time (°C days) in Table 4.3. Thermal time represents the sum of daily mean temperature between zero and maximal rate of maize development (Bonhomme, *et al.* 1994). The method used to compute heat units to estimate thermal time to maturity is described in Appendix A. Except for the moist midaltitude zone, average days to maturity for maize genotypes in the agroclimatic zones fall within the range for recommended varieties (Table 4.3). The fact that farmers in the moist midaltitude zone used later maturing materials (600-series hybrids and lowland locals) rather than materials of intermediate maturity was the reason behind the relatively longer time to maturity reported for maize cultivars in this zone.

It is also clear that the drier the environment (dry midaltitude and dry transitional zones), the faster maize matures and the shorter the effective growing season is in thermal time. More heat units are required for maize to reach maturity in the hot and humid lowland tropics, moist midaltitude zone, and moist transitional zone than in the highland tropics. However, the required heat units accumulate over a shorter time in those zones compared with the highland tropics, where the season is longest in calendar days (Table 4.3).

All farmers, especially those who double-cropped maize, planted their major maize crop in 1992 within or earlier than the average recommended time (Table 4.4). Farmers' planting dates most often coincide with the date at which 80 mm of rainfall accumulate (a better indicator of the beginning of the season) (Table 4.4). The data also suggest a relationship between reliability of rainfall and early planting, as the rate of early planting is much lower in areas where rainfall is highly variable, such as the dry midaltitude and dry transitional zones (Table 4.4). In addition to the effect of the time of onset and reliability of rainfall, there is a considerable variation between zones, farm sizes, and cropping intensity groups in terms of planting time and strategy. For example, the probability of double-cropping increases with diminishing farm size (Table 4.4). An analytical model is developed in the next section to examine the importance of these factors and others in explaining farmers' planting strategies.

Analytical Model of Farmers' Planting Strategy

An analytical model was developed to examine the determinants of farmers' planting choices. The model is based on the assumption that three decisions – how many maize crops to grow each year, which variety to plant, and when to plant – characterize the planting strategy of maize farmers.

Decision 1: Number of maize crops per year

Several factors influence the decision to grow one or two maize crops per year. The physical environment in which maize is grown is the most important factor: to a large extent, climatic attributes such as rainfall patterns and temperature dictate the optimal cropping intensity. Bimodal rainfall is essential for double-cropping, and temperature influences the rate at which maize develops and hence the length of the growing season or production cycle. A second factor, human population density, is an important determinant of the total demand for food and scarcity of farm land and consequently influences the need for more than one maize crop per year. Third, larger families require more maize, although the ability to meet family demand for maize also depends on the total farm area owned or operated by the household. The cropping system is another factor that is important to farmers' planting strategy. Planting maize in association with other crops to meet demands for other farm products, whether for household consumption or for sale (beans, potatoes, groundnuts, etc.), may influence farmers' decision to double-crop maize. Finally, a farmer's age, gender, and level of education may also influence the decision to produce two maize crops per year.

The mathematical specification of the model used to measure and test hypotheses about these relationships is given in equations (4.1) and (4.2):

$$\text{PROSCND} = 1/(1+e^{-z}) \text{ , and} \quad (4.1)$$

$$Z = X'\beta, \quad (4.2)$$

where PROSCND is the probability of a second maize crop, X and β are vectors of independent variables and model parameters, respectively, and e is the base of the natural logarithm. Equations (4.1) and (4.2) define a logistic regression model, which estimates the probability that an event will occur (Maddala 1983). Logistic regression is used to handle cases of dependent variables with a limited range of values, such as whether a farmer will (yes = 1) or will not (no = 0) plant a second maize crop.[5] As is clear from equation (4.1), the logistic model is nonlinear; hence its parameters are estimated using an iterative maximum-likelihood estimation procedure.

The following regressors are contained in the vector of independent variables *X* of (4.2):

- The six final agroclimatic zones (Table 4.3), which reflect the range of climatic conditions (rainfall and temperature variability) under which maize is produced.
- Population density (persons km^{-2}) for the survey sites (1989 census data).
- The ratio of family size to farm size (number of family members per hectare of farm land). This variable measures the combined effects of family size and available farm land.
- The cropping system (1 = pure stand of maize, 0 = intercropped maize).
- The gender and age of the farmer.
- Extension advice on maize variety (1 = farmers who received some extension advice, 0 = farmers who never received extension advice).

Decision 2: What variety to plant?

When farmers are deciding which maize cultivar to plant in the first season (which begins with the March rains), they select the variety that has the desired time to maturity, which in turn is related to whether they plan to produce a second maize crop. As noted earlier, time to maturity is determined largely by moisture and temperature regimes. This model does not attempt to explain farmers' choice of variety, as this decision is influenced by many factors that may be more important than maturity – for example, yield, taste, and tolerance to biotic stresses. Nevertheless, the time it takes a variety to flower and become ready for harvest *is* an important factor in farmers' planting decisions. Average time to maturity (MGTURTY), measured in days between planting and harvesting the maize crop in the first season, is used to represent this decision variable. The relationship between maize maturity and its determinants is specified in equation (4.3):

$$\text{MGTURTY} = f(M,k) + m, \qquad (4.3)$$

where M and k are vectors of predictor variables and model parameters, respectively, and m is the random error term. The elements of M consist of the same regressors used to estimate equations (4.1) and (4.2). The only difference is that the binary variable indicating whether the farmer plans to plant a second maize crop (SCNDCRP) replaces population density in equation (4.3).

Decision 3: When to plant?

After deciding on the intended cropping intensity and the variety to plant during the first season, the farmer decides on the optimal sowing date. This decision variable is measured in days from the beginning of the year (MGSOWDT). Equation (4.4) specifies the model used to analyze the determinants of the sowing date in the first maize season:

$$\text{MGSOWDT} = g(S,a) + u, \tag{4.4}$$

where S and a are vectors of predictor variables and model parameters, respectively, and u is the random error term. To explain variation in the first-season sowing date, the following factors are added to the previous set of regressors:

- Time of onset of the first-season rains, indicating the start of the season. This variable is measured as days from the beginning of the year.
- Sowing method. Sowing maize mechanically, or using animal traction, or manually may influence earliness or timely planting (Waddington *et al.* 1991; Shumba *et al.* 1992).
- Access to machinery services. This variable controls for the effect of ownership of a tractor or oxen on timeliness of planting maize.
- The distance between the maize field and homestead.

The assumptions of the general linear regression model apply to equations (4.3) and (4.4), because all regressors are not correlated with the error terms of the two equations. Accordingly, Ordinary Least Squares (OLS) will yield the best linear, unbiased estimators (BLUE) of the parameters of models 4.3 and 4.4. However, because decisions of *what* and *when* to plant are made jointly, simultaneity and contemporaneous correlation of error terms across equations are assumed. Therefore, the Seemingly Unrelated Regressions Estimation (SURE) procedure (Zellner 1962) is used to estimate the two models as one system of simultaneous equations.

Results of the regression analysis

The three models performed well statistically, as indicated by the goodness of fit and the explanatory powers of the fitted equations (Table 4.5). Table 4.5 reports coefficient estimates and presents estimates of the factor by which the odds of having a second crop [EXP (β)] change as the value of X changes. The odds factor is computed as:

$$\begin{aligned} F_i &= [\text{Prob (SCNDCRP)}] \ \% \ [\text{Prob (No SCNDCRP)}] \\ &= e^z = \underset{e}{\textstyle\sum_i} \beta_i X_i. \end{aligned} \tag{4.5}$$

When the value of β_i is zero (hence Z is zero), the factor F_i equals one, which means the odds do not change when the value of X_i changes (i.e., X_i is not an important determinant of the probability that farmers will plant a second crop). For positive β_i, the factor F_i will be larger than 1 and the odds of the farmer planting a second maize crop increase. The opposite is true for negative values of β_i.

Results show that agroclimate variability, human population pressure, ratio of family size to farm land, and extension advice are important determinants of the probability of double-cropping. The highest increase in the odds [Exp (β)] that a farmer will grow a second crop of maize is associated with the dry transitional and moist midaltitude zones, areas where population density is high, more men than women grow maize, and farmers are older

Table 4.5. Farmers' planting strategies: results of the regression analysis.

	PROSCND (logistic regression)		MGTURTY (days)		MGSOWDT (days)	
Regressors	β	Exp (β)	OLSE of *k*	SURE of *k*	OLSE of *a*	SURE of *a*
Agroclimatic zone:						
Dry midaltitude	−0.10	0.91	−19.36***	−19.46***	–	–
Moist midaltitude zone	0.34**	1.41	7.77**	8.97***	–	–
Dry transitional zone	2.53***	12.61	−25.51***	−26.92***	–	–
Moist transitional zone	−0.77***	0.46	17.82***	18.24***	–	–
Highland tropics	−1.92***	0.15	59.33***	59.13	–	–
Population density (pers. km^{-2})	0.004***	1.00	–	–	−0.01**	−0.01**
Ratio[a]	0.026**	1.03	–	–	−0.24*	−0.25*
PURSTAND MAIZE	−0.15	0.86	1.33	1.28	0.65	0.64
Farmer's age (years)	0.003	1.00	0.24***	0.24***	−0.04	−0.04
Farmer's sex (male)	0.10	1.11	−1.32	1.28	−1.02	−1.04
DISTANCE (home plot)	–	–	–	–	−1.31	−1.16
Extension advice (received)	−0.22***	0.80	5.81***	5.73***	−0.92	−0.84
SCNDCRP (double-cropping)	–	–	−18.92***	−19.01***	0.49	0.68
MGTURTY (days to maturity)	–	–	–	–	–	–
Mechanical planting	–	–	–	–	−4.09***	−4.28***
Use own oxen or tractor	–	–	–	–	−3.42**	−3.53**
Onset of rains (days)	–	–	–	–	0.76***	0.78***
Constant	−1.44**	–	146.40***	146.12***	42.06**	40.47***
	χ^2	228***	R^2	0.51	R^2	0.27
	n	921		0.36	*F*	27.07***
			F	82.98***	*n*	806
			n	806	System weighted R^2 = 0.42	
			System weighted R^2 = 0.42			

Note: *, **, and *** indicate significance at the 10%, 5%, and 1% levels, respectively. PROSCND = probability of a second maize crop; MGTURTY = average time to maturity; MGSOWDT = farmer's optimal sowing date; OLSE = Ordinary Least Squares Estimates; pers. = persons; SURE = Seemingly Unrelated Regressions Estimates.

[a] Ratio of family size to farm size (number of family members per hectare of land).

(nonnegative values of β in Table 4.5). Contact with extension seems to reduce the chances of a second maize crop, which may imply that extension advice tends to promote single maize crops or may result from mere statistical correlation indicating that extension services are more concentrated in areas where single-season maize dominates. Single-season maize is also associated with large-scale commercial maize farming in high-potential areas (Tables 4.3 and 4.4). This result could suggest an undesirable bias in extension services towards large-scale farmers and late-maturing germplasm for single-cropping in high-potential lands.

Parameter estimates of the logistic regression reported in Table 4.5 are used to compute the probability of farmers planting two maize crops in the various agroclimates. The estimated logistic equation was evaluated at average sample levels of the independent variables and gave good predictions of the probability of a second maize crop (Table 4.6). When the effects of population density, extension advice, and other factors were controlled for in model 4.1, the probability of double-cropping in the dry transitional zone became very high (0.95) compared with the rate of double-cropping observed from the survey (0.76; see Table 4.3). At the same time, the model predicted a lower probability of a second maize planting in the dry midaltitude and highland tropical zones than observed from the survey (Tables 4.3 and 4.6).

What and when to plant (Models 4.3 and 4.4)

Systems estimation (SURE) of models 4.3 and 4.4 did not significantly alter the values and statistical significance of the OLS parameter estimates (Table 4.5). While this indicates that contemporaneous correlation of error terms across equations was weak in the estimated systems, SURE estimates are nevertheless more efficient than OLS estimators in large samples (Johnston 1984). As expected, average weeks to maturity were highest in the highland tropics, where maize took 205 days (146 + 59) to mature (Table 4.5). On the other hand, the shortest time to maturity was found in the lowland tropics (the coast), where maize matured in 106 days (146 − 40) on average.[6] The average maturity in the moist transitional zone was shorter than in the typical highland tropical areas and longer than in the typical midaltitude areas. Farmers attempting a second crop (SCNDCRP) used maize germplasm that reaches maturity five weeks earlier than single-season maize (18.9 × 2). Maize planted in pure stand matured later, on average, than intercropped maize. The effect of the gender of the farmer was not statistically significant, suggesting that male and female farmers make similar decisions. The farmer's age was an important determinant of maize maturity: older farmers tended to select maize genotypes with longer time to maturity. Farmers who received extension advice tended to plant earlier and use maize materials that took relatively longer to mature (Table 4.5). These results agreed with the earlier findings from the logistic regression analysis that suggested that contact with extension was associated with planting medium- to late-maturing maize in high-potential areas.

In magnitude and level of statistical significance, sowing method and

Table 4.6. Simulation of farmers' planting strategy by agroclimatic zone, Kenya.

Agroclimatic zone	Population density (pers. km^{-2})	Average age of farmer (years)	Average ratio (pers. ha^{-1})	Percentage received extension advice	Percentage farmers growing maize in pure stand	Percentage area pure stand maize	Percentage male farmers	Probability of a second maize crop	Average maturity (days)
Lowland tropics	121	47.2	6.1	41	29	46	67	0.31	116
Dry midaltitude zone	211	46.5	3.4	48	21	53	55	0.37	139
Moist midaltitude zone	310	48.4	5.0	57	30	57	54	0.58	146
Dry transitional zone	398	47.5	5.8	55	15	60	63	0.95	147
Moist transitional zone	331	45.0	6.4	77	25	96	59	0.33	162
Highland tropics	238	44.8	5.0	59	20	93	58	0.10	196

Note: predicted using parameter estimates of equations (4.1–4.4) and survey data.
pers., Persons.

access to/ownership of mechanical means of sowing were the most important sources of variation in the date of sowing first-season maize. The next most important sources of variation were the onset of the rains and population pressure (Table 4.5). Farmers who used mechanical means of sowing maize – tractors or oxen – planted four days earlier on average than farmers who sowed maize by hand. The term controlling for the interaction between ownership of tractors or oxen and mechanical sowing ('use own') indicates that farmers who used their own tractors or oxen planted earlier than farmers who hired them (i.e., used but did not own). The high statistical significance and magnitudes of the regressors of all three equations indicate the importance of these factors to the planting strategy of Kenyan maize farmers.

Major Characteristics of Maize Farming Systems in Kenya

Clearly, in addition to physical climate, many other variables influence maize cropping patterns and intensity, and consequently determine maturity requirements and farmers' planting strategy and varietal selection. These include human population pressure, available farm land, family size, and food supply strategies of the farming household. All of the parameters shaping maize farming in the various agroclimatic zones provide important information on system constraints that should be considered when researchers develop technologies for farmers. Farmers' objectives should be different in areas where population pressure is high compared with areas where land is abundant. For instance, when a large family cultivates a very small land holding, the family's major concern is to increase land productivity to produce enough food for the family and some surplus for sale. In these circumstances, farming practices that increase output per unit of area (e.g., double-cropping or multiple-cropping) and use more labor are adopted. Moreover, farmers who are better integrated in the market economy and have better access to trade opportunities tend to specialize most often in commercial crops, which they can sell for cash to meet household needs. These factors are the main reason why multiple-cropping dominates smallholder subsistence agriculture in Africa (Beets 1982), and they have important implications for research, because technologies and recommendations developed and tested under monocultural systems do not always suit multiple-cropping systems. Another source of variation in the objectives of different groups of farmers is gender: male farmers may differ from female farmers in their assessment and hence adoption of various production technologies. Identifying the major characteristics and parameters of maize farming in Kenya and understanding their implications for maize research and technology design enables scientists to fit the products of research more closely to the circumstances of different groups of farmers. The sections that follow describe the salient characteristics of maize farming in Kenya (listed in Table 4.7).

Population density and land pressure

Further evidence is provided in Table 4.7 for the relative importance of population density and its effects on maize farming in Kenya. The data indicate strong associations among the density of human population, average farm size, intensity of maize cultivation, multiple cropping, percentage of land devoted to maize production (for food), and pressure on the natural resource base. Where population density is high, as in the dry transitional zone, most maize is produced on very small holdings (1.5 ha on average) and double-cropping, intercropping, and the pressure on erosion-prone sloping land are high (Table 4.7). Although the same is true of the moist midaltitude areas where rainfall is bimodal, double-cropping is practiced less in the moist transitional and highland tropical zones, where the growing season is relatively longer. Even so, many small-scale farmers in these zones attempt to produce two maize crops every year, and nearly all of these farmers intercrop maize with beans (Table 4.7). It is also important to note that, although the average farm size is larger in the moist transitional zone, the percentage of farmers cultivating less than 2 ha is higher than in the moist midaltitude zone. This suggests that land distribution is more skewed in the moist transitional and highland tropical zones than in the moist midaltitude zone. Moreover, the share of maize in total farm area declines with increasing elevation and rainfall, whereas the percentage of maize grown on sloping to very steep land increases with elevation, as expected (Table 4.7).

All of these relationships point to the importance of considering population density and the quantity and quality of available land when designing maize technologies, to ensure that they are efficient and protect resources. For example, cultivars with a shorter time to maturity, better tolerance of low soil fertility, and reduced dependence on external inputs will be appropriate for helping resource-poor maize farmers in the dry and moist portions of the midaltitude and transitional zones. On the other hand, maize research needs to concentrate on developing technologies that increase the productivity of capital and labor in high-potential areas where medium- to large-scale maize farming is most dominant.

Cropping system

The vast majority of small-scale farmers in all zones, as well as many large-scale farmers in the midaltitude and lowland tropical zones, intercrop maize with beans or cowpeas (Table 4.7). Any maize technology should therefore be tested under intercropping conditions before being introduced. For example, it will be important to evaluate how the new technology affects competition between the intercrops above and below the soil surface, and to understand the implications for crop nutrient needs, planting strategies, plant spacing, and weed and insect control. The KARI maize breeding program and farming systems research should focus on identifying and selecting traits that are desirable in maize varieties grown with other crops.

Table 4.7. Maize farming systems parameters by agroclimatic zone, Kenya.

Agroclimatic zone	Percentage very small farms (< 2 ha)	Average farm size (ha)	Percentage land under maize		Other enterprises in pure stand	Percentage farmers intercropping maize		Major intercrop	Percentage maize on slopes and very steep land	Most serious constraints (% crop loss)[a]	Average age of farmer (yr)	Female farmers (%)	Farmers with no education (%)
			Small (< 8 ha)	Large (> 8 ha)		Small (< 8 ha)	Large (> 8 ha)						
Lowland tropics	60	2.4	66	7	Beans, roots, and pasture	78	50	Cowpeas/ cassava	49	RAIN, STBR, OPST, OWED	47	33	75
Dry midaltitude zone	61	3.6	64	31	Beans, pasture, and roots	88	77	Beans, cowpeas	58	RAIN, OWED, CHGB, OPST	47	45	43
Moist midaltitude zone	56	2.3	38	12	Beans, pasture	77	50	Beans	57	STRG, RAIN, FRTL, OWED	48	46	29
Dry transitional zone	80	1.5	57	–	Beans, pasture	95	–	Beans	85	SMUT, RAIN, OWED, CHGB	48	37	31
Moist transitional zone	64	7.7	46	26	Pasture, beans, and potatoes	89	16	Beans	67	STBR, LOGN, HAIL, RAIN	45	41	39
Highland tropics	54	10.9	45	12	Pasture and wheat	90	39	Beans	65	RAIN, HAIL, STBR, LOGN	45	42	40
Total	60	7.2	–	–	–	–	–	Beans/ cowpeas	64	–	46	41	41

[a] Constraints include drought and variable rainfall (RAIN); stalk borer (STBR), chafer grubs (CHGB), and other insect pests (OPST); *Striga* (STRG) and other weeds (OWED); head smut diseases (SMUT); low soil fertility (FRTL); hail (HAIL); and lodging (LOGN).

Major stress factors and productivity constraints

Farmers were asked to name (rather than rank) the major problems facing maize production in their fields. Farmers' estimates of quantitative aspects of the problems they identified were then used to develop a more objective assessment and ranking of those problems. The percentage of land affected by the problem, average percentage damage caused, and frequency of the problem were used to estimate the average percentage crop loss per year resulting from each problem.

Farmers identified inadequate rainfall (drought), variability of rainfall, pests, diseases, weeds, and low soil fertility as the most serious stress factors in maize production in Kenya (Table 4.7). However, the relative importance of these biotic and abiotic factors varies among the six agroclimatic zones (Plate 4), indicating that the zones do represent different biotic and abiotic stress environments. These results confirm the findings of the earlier analysis of germplasm adaptability and maturity requirements, which identified the six zones as distinct maize production environments and target domains for developing maize technologies.

Relatively high average age of farmers

The average age of farmers is relatively high in all zones (Table 4.7), indicating the age bias in access to farm land. Younger members of the rural community in maize-producing areas work as hired hands or provide labor on the family farm or in urban areas for relatively long periods before acquiring their own land. However, the high average age of farmers implies that there is a stable land tenure system in maize farming and that maize farms are basically operated as family firms. Because maize technologies are used and evaluated by relatively older farmers, who are generally more experienced but are expected to be more averse to risk and resistant to change, farmers' age should be considered when planning maize research and extension. A suitable mechanism is also needed to deliver extension messages according to age group.

The gender factor

More than 40% of the sampled maize farmers were females (Table 4.7). The fact that so many users of maize technologies are women indicates the importance of considering the special needs of these farmers in designing maize technologies and in extension work.

Maize cultivation on slopes

A considerable number of farmers (64%) cultivate maize on sloping to very steep plots (Table 4.7). Maize research must place special emphasis on developing and testing technologies that will help mitigate erosion and conserve soils.

Schooling

Except in the lowland tropics, many farmers reported having received some formal or informal schooling. Education has been found to enhance farmers' ability to acquire and understand information about new technology and to apply modern inputs efficiently (Huffman 1974). This factor is important for targeting maize farmers in the lowland tropics.

Conclusion: Implications for Research

The analyses described above show that farmers' practices in the maize adaptation zones currently used by the national maize research program deviate significantly from research recommendations in terms of germplasm suitability and maturity. This divergence was highest in the midaltitude zone, where farmers tend to plant maize hybrids that mature later than the medium-maturing materials developed for this environment. New boundary conditions were developed and zones reclassified to distinguish a new transitional zone between the typical midaltitude and highland tropical zones. Areas encompassed by the new zone were slightly cooler than areas belonging to the midaltitude zone and warmer than those in the highland tropics.

Revising the classification of the agroclimatic zones led to a better characterization of maize production environments. Nevertheless, even after the relatively cooler segments of the midaltitude zone had been reclassified as part of the transitional zone, the survey data indicated that farmers remaining in the midaltitude zone – especially the moist areas – were using late-maturing materials rather than the medium-maturing hybrids researchers recommended for that zone. Farmers used later maturing materials even though the high population density and bimodal rainfall of the zone would favor double-cropping and genotypes that mature more rapidly.

A historical bias toward farmers in the highland tropics was observed in KARI's and the Kenya Seed Company's efforts to develop and distribute maize varieties. National maize research strategies and resource allocation should shift to address the historical underinvestment in the midaltitude and transitional environments, where most of Kenya's maize area and farmers are located.

The fact that maize farmers in the transitional zone have distinct germplasm needs was confirmed by the logistic and multiple regression analyses of the determinants of farmers' planting strategies. A higher probability of growing two maize crops per year is associated with higher human population density, bimodal rain patterns, a higher ratio of family members to land, and poorer access to extension advice. On average, maize materials planted in the transitional zone matured earlier than cultivars grown in the highland tropics and later than cultivars planted in the midaltitude zones. All of these results have strategic implications for planning maize germplasm research in Kenya, which has not adequately catered to the needs of farmers in the transitional zone, who cultivate about 45% of Kenya's maize area. The highland tropical program at the National Plant Breeding Research Center in Kitale screens for late maturity and discards materials that mature in less than 180 days, whereas the Regional Research Center in Embu selects materials for medium maturity (120–150 days to maturity) and discards materials maturing in more than 150 days. These practices exclude materials maturing in 150–180 days, the range that survey results indicate is desirable for the transitional zone (Table 4.3).

It is of prime importance that maize researchers begin screening materials suitable for this transitional zone. Recently (in 1991) a maize breeder was posted to the Regional Research Center in Kakamega, in the middle of the transitional zone. However, only maize germplasm at a late phase of development (the national performance trials) is tested at Kakamega (Njoroge *et al.* 1992a), and the Regional Research Center should become involved in maize breeding at much earlier stages (testing top crosses, evaluating progeny, and conducting preliminary yield trials).

Although the term 'transitional' is used to indicate something less than a distinct agroclimatic zone, there is increasing evidence that the transitional zone does represent a distinct biophysical environment for maize production. For example, differences in the disease spectrum in true midaltitude and cooler transitional environments have become more evident. These differences have been recognized by the CIMMYT maize breeding program for some time (CIMMYT 1988).

The information used to develop a better delineation of Kenya's maize agroclimatic zones has also enabled the major characteristics of maize farming in Kenya to be defined more clearly. The most important system characteristics, constraints, and stress factors that shape maize production and require special attention in planning maize research and designing technology are:

- The dominance of intercropping, particularly of maize and beans.
- The high percentage of maize farmers cultivating very small plots and attempting two maize crops every year.
- The significant influence of population density on intensity of maize cultivation (double-cropping), the time to maturity of the germplasm farmers select, and pressure on fragile lands (steep slopes).
- The variation in major biotic and abiotic stress factors for maize production across the six zones.
- The high percentage of women maize producers.
- The significant proportion of the maize crop sown on sloping to very steep plots.
- The relatively high age of maize farmers.

The first three factors require a research strategy focusing on technology to raise land productivity and intensify labor use. In areas of high population pressure and bimodal rainfall, cultivars that require less time to mature (to allow two crops per year and a better spread of labor use over time) will be desired. Other traits in maize materials and production methods that are suitable for multiple cropping will also be needed for increased productivity of the land in mixed cropping areas. On the other hand, increased productivity of capital and labor is needed for large-scale commercial maize farmers in Kenya's high-potential areas. Higher response to external inputs (chemical fertilizer, pesticides, herbicides, etc.) and increased efficiency of their use (levels and timing) should be the main objectives of maize research for this target group.

It is also important to target the six maize agroclimates as distinct zones when designing future technological interventions for maize in Kenya. The fact that a significant proportion of the maize research clients and adopters of the technologies are women and relatively older farmers needs to be taken into consideration in planning maize research and developing technologies. Also, since a considerable share of the land under maize is sloping to very steep, maize technologies need to be tested under similar conditions for erosion and soil conservation effects.

The findings described here demonstrate how the characterization of maize production systems and environments was improved significantly by pairing survey data on farmers' practices and priorities and other socioeconomic factors with agroclimate attributes. This exercise confirms the importance of melding these kinds of data for properly defining crop production environments and accurately targeting research.

References

Beets, W. (1982) *Multiple Cropping and Tropical Farming Systems.* Westview Press, Boulder, Colorado.

Bonhomme, R., Derieux, M., and Edmeades, G. (1994) Flowering of diverse maize cultivars in relation to temperature and photoperiod in multilocation field trials. *Crop Science* 34, 156–164.

CIMMYT (1988) *Maize Production Regions in Developing Countries.* International Maize and Wheat Improvement Center (CIMMYT), Mexico City.

Gebrekidan, B, Wafula, B., and Njoroge, K. (1992) Agroecological zoning in relation to maize research priorities in Kenya. In: *KARI: Review of the National Maize Research Program.* Proceedings of a Workshop Held between November 19 and 23, 1991, Kakamega, Kenya. Kenya Agricultural Research Institute (KARI) and the International Service for National Agricultural Research (ISNAR), Nairobi, Kenya.

Huffman, W. (1974) Decision making: The role of education. *American Journal of Agricultural Economics* 55, 85–97.

Johnston, J. (1984) *Econometric Methods.* McGraw-Hill, New York.

KARI (Kenya Agricultural Research Institute) (1992) *Improve and Sustain Maize Production through Adoption of Known Technologies.* Information Bulletin No. 7. KARI, Nairobi, Kenya.

Maddala, G. (1983) *Limited-dependant and Qualitative Variables in Econometrics.* Cambridge University Press, Cambridge, U.K.

Njoroge, K., Kanampiu, N., Otsyula, R., Muthamia, Z., Gathuri, C., and Chivatsi, W. (1992a) The high altitude maize breeding program. In: *KARI, Review of the National Maize Research Program.* Proceedings of a Workshop held between November 19 and 23, 1991, Kakamega, Kenya. Kenya Agricultural Research Institute (KARI) and the International Service for National Agricultural Research (ISNAR), Nairobi, Kenya.

Njoroge, K., Kanampiu, N., Otsyula, R., Muthamia, Z., Gathuri, C., and Chivatsi. W., (1992b) The Coast maize improvement program. In: *KARI, Review of the National Maize Research Program.* Proceedings of a Workshop held between November 19 and 23, 1991, Kakamega, Kenya. Kenya Agricultural Research Institute (KARI) and the International Service for National Agricultural Research (ISNAR), Nairobi, Kenya.

Norusis, M. (1991) *SPSS/PC+: Advanced Statistics 4.0 for the IBM PC/XT/AT and PS/2.* SPSS, Chicago, Illinois.

Otichillo, W.K., and Sinange, R.K. (1991) *Long Rains Maize and Wheat Production in Kenya in 1990.* Technical Report No. 140. Department of Resource Surveys and Remote Sensing, Ministry of Planning and National Development, Nairobi, Kenya.

Ransom, J., and Osoro, M. (1991) Effects of planting date and genotype on *Striga hermonthica* parasitism on maize and sorghum in Kenya. Poster presented at the 22nd International Plant Protection Congress, 11–16 August, 1991, Rio de Janeiro, Brazil.

Shumba, E.M., Waddington, S., and Rukuni, M. (1992) Use of tine-tillage, with Atrazine weed control, to permit earlier planting of maize by small holder farmers in Zimbabwe. *Experimental Agriculture* 28, 443-452.

Waddington, S., Mudhara, M. Hlatshwayo, M., and Kunjeku, P. (1991) Extent and causes of low yield in maize planted late by smallholder farmers in subhumid areas of Zimbabwe. *Farming Systems Bulletin, Eastern and Southern Africa* 9, 15–31.

Zellner, A. (1962) An efficient method of estimating seemingly unrelated regressions and tests of aggregation bias. *Journal of the American Statistical Association* 57, 348–368.

Endnotes

1 The agroclimatic zones initially developed by the MDBP (Chapter 2) were consistent with the maize adaptation zones used to organize maize research in Kenya. Although the initial scheme separated dry from moist areas in all zones, it did not contain a transitional environment for maize production (Plate 1). This chapter, however, provides evidence that supports the need for such an environment and a revision of the initial zones.

2 A varietal hybrid is a 'nonconventional hybrid' derived from OPVs that are genetically different enough to express hybrid vigor (unlike 'conventional hybrids,' which are obtained from inbred lines).

3 This could also be true for maize that farmers identify as Katumani Composite, because most farmers do not buy new seed to replace the certified variety for many years.

4 Note that the local cultivars popular in the lake region differ from the local cultivars that are popular near the coast, though all of these materials are lowland materials.

5 When the dependent variable can have only two values (binary), the assumptions of the standard linear regression model about the distribution of the random error component, such as multivariate normality and independence of regressors and error terms, are necessarily violated (Maddala 1983; Norusis 1991).

6 To avoid singularity of the $x'x$ matrix, one dummy is dropped using the rule ($\Sigma_i d_i = 1$, where d_i is the deviation of the mean of category i from the overall mean, and $i = 1, 2, \ldots N-1$, where N is the total number of dummies. The coefficient for the lowland tropics (Coast) was accordingly derived to be the negative sum of the coefficients of the zone dummies that were included.

II *Ex Ante* Research Evaluation and Priority Setting

5 Relevance of Maize Research in Kenya to Maize Production Problems Perceived by Farmers

Rashid M. Hassan, Ruth Onyango, and J.K. Rutto

Geo-referenced survey data on farmers' practices, data on types, numbers, and sites of trials conducted by the maize research program, data on resources dedicated to maize trials, and aggregate area and population statistics were examined to evaluate the relevance of recent maize research to farmers' needs. Economic efficiency and nonefficiency criteria were used to rank maize production zones in order of their importance for maize research. The moist transitional zone ranked first, followed by the highland tropics and moist midaltitude zones. An analysis of trends in types of maize research and funding over the past 12 years indicated that more than half of the maize research program resources (measured in numbers of experiments) were directed to the highland tropics, whereas the zone with the highest potential to benefit from research – the moist transitional zone – received considerably less attention. Farmers' assessment of problems revealed that more breeding research is needed in the high-priority zones than has been done there in the past. Crop management research received more resources than required in the moist midaltitude zone but less than needed in the dry midaltitude zone. These trends must be reversed or balanced if maize research is to become more responsive to farmers' conditions and if funding of research in high-payoff zones and programs is to be sustained.

This chapter evaluates the relevance of past research on maize in Kenya by identifying areas where research has departed from the needs and priorities described by farmers. The MDBP combined survey information on farmers' perceptions of important constraints to maize production with spatial data on agroclimatic characteristics, area planted to maize, and population density. Economic efficiency and nonefficiency criteria were applied to these data to rank each zone, constraint to maize production, and research program in order of its priority for maize research. This set of priorities is compared with

KARI's historical allocation of maize research resources by zone and research program.

Data and Methods

Agroclimatic zones provide a framework for applying efficiency criteria to assess potential benefits from new crop varieties and improved production methods and to identify zones where the payoff to research is expected to be high. Also, because the distribution of poor farmers and marginal lands is often correlated with variation in the physical environment and the natural resource base (Trainer 1989; Wood and Pardey 1993), agroclimatic zoning provides a better structure for determining how research can meet goals of equity (at least on the supply side) and sustainability.

Data for this study were compiled from several sources:

- The maize-specific agroclimatic zones described in Chapters 2 and 4.
- Data from the geo-referenced survey of maize farmers across 30 districts of Kenya (Chapter 3).
- Records of experiments conducted by the maize research program at KARI over the past 12 years. This information was organized by site, year, and type of trial; source of funding; and experimental factors (for details, see Hassan *et al.* 1992).
- Data on population density and area planted to maize from CBS and DRSRS at the Ministry of Planning and National Development, Kenya.

A scoring method (weighted criteria model) was used to evaluate research priorities by maize agroclimatic zone and production constraint. The scoring method offers a relatively formal, structured, yet simple means of setting research priorities when data limitations prevent other measures from being used. Results of the analysis are more easily interpreted for research managers and decision makers than results of alternative methods, such as economic surplus measures. Alston *et al.* (1995) have concluded that, if applied carefully, the scoring method can generate results consistent with economic principles. However, scoring and other 'short-cut' methods of *ex ante* research evaluation often do not adequately model potential impacts on consumers.

Alston *et al.* (1995) regard two efficiency indices as good approximations of economic surplus measures: the gross efficiency index (G_i) and net efficiency index (N_i). The gross efficiency index for class i (zone, commodity, research program, etc.) is defined as the product of four factors:

$$G_i = P_iQ_i \times Ey_i \times p_i \times T_i. \quad (5.1)$$

The factors are a baseline value of production (P_iQ_i); expected productivity gain (Ey_i), measured as proportional unit cost reduction or yield increase; probability of success (p_i); and the proportion of farmers likely to adopt the technology (T_i). The net efficiency index (N_i) is derived in (5.2) by dividing G_i by research costs (R_i):

$$N_i = G_i/R_i. \tag{5.2}$$

The net efficiency index accounts for research costs, but it assumes the same timing of costs and benefit flows and/or a zero discount rate. Another disadvantage is that the net efficiency index tends to bias the ranking towards smaller (i.e., lower cost) programs (Alston *et al.* 1995).

The gross and net efficiency indices are ordinal measures of the efficiency objective, meaning that they are best used to rank alternatives in order of priority rather than to determine quantitative allocations of resources. After the efficiency ranking is obtained, it is recommended that nonefficiency criteria be applied and used as 'transformations' of the efficiency index (Alston *et al.* 1995).

Determining High-Priority Agroclimatic Zones for Maize Research

Data used to derive baseline maize production parameters for the six agroclimatic zones are presented in Table 5.1. The gross efficiency index (G_i) is calculated in Table 5.2 to rank the zones in order of their contribution to gross maize productivity gains. The following variables were used to compute G_i:

1. Total area planted to maize for the major maize production season of 1990, defined by the March (long) rains (Otichillo and Sinange 1991).
2. Average yield of maize planted during the March rains, calculated from survey farmers' estimates of three yield variables: yield during the survey year (1992), yield obtained in the previous year, and yield in a 'normal' (average) year.
3. Area sown to maize in the season defined by the October (short) rains, computed from survey data as the proportion of the total maize area planted during the March rains that was planted to maize during the second season (short rains).
4. Average yield during the short rains, calculated as in (2) above.

These variables were used to compute total realized or baseline production of maize (Q_i) (Table 5.1). A ranking was derived on the basis of zonal shares in Q_i – that is, realized rather than potential production gains.

5. Feasible proportional yield increase (Ey_i). Long-term data from maize trials (March rains) were used to define potential maize yield. However, as yields obtained in researchers' trials do not reflect the yield potential of maize grown in farmers' fields, the feasible proportional yield increase was calculated as 50% of the research–farmer yield gap (Table 5.2).
6. Probability of success (p_i). Information was not gathered from researchers and other sources on the probability of achieving the feasible yield increase. As an alternative, the proportion of potential (research) yield realized by farmers was used to indicate the probability of success.
7. Percentage of adopters (T_i). This variable was approximated by the percentage of farmers who had ever bought improved seed. Information on the

Table 5.1. Maize production by agroclimatic zone and season, Kenya.

	Major season		Proportion of major-season maize area	Minor season			
Agroclimatic zone	Maize area (000 ha)[a]	Maize yield (t ha^{-1})[b]	planted to maize in the second season[a]	Maize yield (t ha^{-1})[b]	Maize production[c] (000 t)	(%)	Priority ranking[d]
Lowland tropics	33	1.36	0.25	0.99	53	(2)	6
Midaltitude zone							
Dry	118	1.03	0.41	0.83	162	(6)	4
Moist	118	1.44	0.47	1.11	232	(9)	3
Transitional zone							
Dry	37	1.21	0.79	1.08	76	(3)	5
Moist	424	2.76	0.10	1.50	1234	(46)	1
Highland tropics	307	2.91	0.03	1.73	909	(34)	2
Total	1037	2.31	0.20	1.33	2666	(100)	–

Source: MDBP.
Note: the major growing season for maize occurs during the long rains, which extend from March to August. The minor season occurs during the short rains, from October to February.
[a] Area estimates based on Otichillo and Sinange (1991).
[b] Yield is the average of three observations made by farmers about their maize yields: yield during the survey year (1992), yield in the preceding year (1991), and yield in a 'normal' year.
[c] Total major and minor season production.
[d] 1 = most important, 6 = least important zone.

history of adoption was collected from farmers through retrospective questioning.
8. The maize price (P_i). A simple average of maize prices (1989–1993) was calculated from the Ministry of Planning and National Development time-series on market commodity prices (Mills *et al.* 1994). These data show considerable spatial variation, reflecting mainly transportation costs (Table 5.2).

Because data on research costs were not available, the net efficiency index (N_i) was not derived. Two values were computed for G_i: $G1_i$ and $G2_i$. The $G2_i$ derives the efficiency index after accounting for spillover effects. The proportion of farmers from other zones using varieties developed for the zone in question was used as a proxy for spillover effects: the higher the percentage of farmers in nontarget zones using target environment varieties, the bigger the spillover effect (Table 5.3). Maize varieties and target zones included (KARI 1992):

- Coast Composite and Pwani Hybrid for the lowland tropics.
- Katumani and Makueni Composites for the dry midaltitude zone.

Table 5.2. Agroclimatic zones for maize production in Kenya, ranked by gross efficiency criteria.

Agroclimatic zone	Maize production (000 t)[a] Q_i	Yield potential (t ha^{-1})[b]	Possible proportional yield increase[c] Ey_i	Proportion of potential yield realized[d] p_i	Proportion of farmers ever bought improved seed T_i	Maize prices[e] (ratio) P_i	Gross efficiency index[f] $G1_i$
Lowland tropics	53	3.11	0.64	0.44	0.43	1.0	6.4 (6)
Midaltitude zone							
Dry	162	2.67	0.84	0.39	0.67	0.78	27.7 (4)
Moist	232	4.01	0.89	0.36	0.79	0.72	42.3 (3)
Transitional zone							
Dry	76	3.34	0.88	0.36	0.91	0.78	17.1 (5)
Moist	1234	7.12	0.79	0.36	0.98	0.63	216.7 (1)
Highland tropics	909	7.71	0.83	0.38	0.98	0.65	182.6 (2)

Source: MDBP.

[a] From Table 5.1.

[b] Long-term yield data from the National Performance Yield Trial (NPT) sites (KARI 1992).

[c] The feasible proportional yield increase is estimated to be 50% of the difference between yields obtained in research trials (potential yields) and by farmers for the March–August season.

[d] The ratio of farmers' average yields to potential yields.

[e] The ratio of the 1989–1993 average price to the price at Mombasa (border FOB price) over the same period (Mills *et al.* 1994).

[f] $G1_i = Q_i \times Ey_i \times p_i \times T_i \times$ Ratio.

- Varietal hybrids H511 and H512 for the moist midaltitude and dry transitional zones.
- Varietal hybrids H614, H622, and H632 for the moist transitional zone.
- H614 and conventional hybrids H625 and H626 for the highland tropics.

Before spillover effects were accounted for, the moist transitional zone ranked top in priority for research, followed by the highland tropics and then the moist midaltitude zone. After spillover effects were accounted for, the $G1_i$ rankings were reversed for the dry transitional and dry midaltitude zones (Table 5.3), indicating that the suitability of the drought-tolerant varieties targeted for the dry midaltitude zone is more limited than that of cultivars recommended for the dry transitional zone (the 500-series hybrids).

Nonefficiency criteria were applied to the G_i ranking in Table 5.3, based on objectives other than growth and efficiency:

- Reduce relative inequality among producers – in other words, assign a higher weight to redistributing benefits of research to resource-poor

Table 5.3. The contribution of maize zones to efficiency, equity, and sustainability objectives, Kenya.

Agroclimatic zone	Gross efficiency index $G1_i$	Spillover effect[a] s_i	Gross efficiency index[b] $G2_i$	Population density[c] (pers./000 km²)	Population density[c] (density factor) m_i	Gross efficiency-equity index[d] $G3_i$	Proportion of farmers for whom erosion is serious problem r_i	Gross efficiency-equity-sustainability index[e] $G4_i$	Maize experiments, 1980–92[f] (no.)	Maize experiments, 1980–92[f] (%)	Rank
Lowland tropics	6.4(6)	0.01	0.1(6)	121	0.31	0.03(6)	0.23	0.01(6)	93	12.2	(4)
Midaltitude zone											
Dry	27.7(4)	0.13	3.6(5)	210	0.53	1.9(5)	0.48	0.91(5)	105	13.8	(3)
Moist	42.3(3)	0.22	9.3(3)	310	0.78	7.3(3)	0.34	2.5(3)	14	1.8	(6)
Transitional zone											
Dry	17.1(5)	0.22	3.8(4)	398	1.00	3.8(4)	0.56	2.1(4)	42	5.5	(5)
Moist	216.7(1)	0.28	60.7(1)	331	0.83	50.4(1)	0.32	16.1(1)	142	18.6	(2)
Highland tropics	182.6(2)	0.29	53.0(2)	238	0.60	31.8(2)	0.38	12.2(2)	367	48.1	(1)
Midaltitude zone (previous classification)	–	–	–	–	–	–	–	(1)	198	25.9	(2)

Source: MDBP.

Note: numbers in parentheses indicate priority ranking of zone (1 = most important).

pers., Persons.

[a] Proportion of farmers in other zones adopting cultivars targeted for this zone.

[b] $G2_i = G1i \times s_i$.

[c] Average of population density in survey sites within the zone (i.e., sample average, according to the 1989 population census. The highest density zone (dry transitional zone) is used as the base to compute the population density factor.

[d] $G3_i = G2_i \times m_i$.

[e] $G4_i = G3_i \times r_i$.

[f] Of 780 maize experiments conducted by KARI between 1980 and 1992, 763 (98%) were 'valid' for this analysis (i.e., the site of the trial could be identified).

[g] The midaltitude zone originally encompassed the moist midaltitude zone and the dry and moist transitional zones (see Chapters 2 and 4).

farmers. This objective was approximated by a criterion assigning higher priority to zones having a higher proportion of small-scale farmers (farmers having less than 2 ha) (Table 5.3).

- Provide an adequate food supply and employment for the rural population. The number of people affected by or likely to benefit from a new maize technology was considered an important criterion for assessing contributions to this objective.

Both of these objectives are closely related, and they were accounted for through the population density criterion, which gave zones with a higher population density a higher priority (Table 5.3).

- Promote sustainable agricultural production and conserve resources. The percentage of farmers who indicated that soil erosion was a serious problem was used to approximate the sustainability criterion. A higher priority was assigned to higher probabilities of soil erosion (Table 5.3).

The distribution and sustainability criteria were applied multiplicatively to the gross efficiency index $G1_i$ to compute $G3_i$ and $G4_i$ (Table 5.3). After equity criteria were taken into account, the higher ranking of the dry transitional zone was further confirmed, basically because of the population effect. The dry transitional zone is the most heavily populated region in Kenya compared with the dry midaltitude zone, which is the second least populated (Table 5.3). On the other hand, the sustainability criterion did not change the $G3_i$ ranking. It did show, however, that the dry transitional and dry midaltitude zones have the highest rates of soil erosion (Table 5.3). If the rankings in Table 5.3 were to guide the allocation of KARI's research resources, the moist transitional zone would be given top priority, followed by the highland tropical and moist midaltitude zones.

Priority Assigned to Agroclimatic Zones by the Maize Research Program

As mentioned earlier, no attempt was made to determine research costs because of data limitations, especially insufficient data on the cost of research by agroclimatic zone. As an alternative, records of experiments at seven major maize research centers over the past 12 years were retrieved[1] to review the relevance of KARI's maize research program by zone and discipline. The total number of experiments done at sites within each agroclimatic zone was used to estimate of the share of maize research resources that KARI allocated to the zone.

Several important differences were observed between the ranking of zones by efficiency/equity criteria and the ranking defined by past research emphasis and resource allocation within KARI (Table 5.3). The highland tropics received top priority for maize research but ranked second according to efficiency/nonefficiency measures. About half of the experiments done by the maize research program over the past 12 years were done in that zone. In terms of efficiency and nonefficiency measures, the highest ranking zone was the midaltitude zone (encompassing the moist midaltitude zone and the dry and moist transitional zones), which received about half the number of experiments conducted in the highland tropical zone between 1980 and 1992. Clearly KARI needs to reallocate its maize research resources away from the highland and lowland tropical zones to give greater emphasis to the transitional and midaltitude zones (Table 5.3).

Farmers' and Researchers' Assessments of Important Maize Production Constraints

Farmers' perceptions of the most important constraints to maize production were obtained from a structured questionnaire and used to prioritize production problems. A number of measures were used to improve the efficiency and reliability of farmers' assessments. Farmers are often asked to rank problems in order of importance or seriousness. For the Kenya survey, farmers were asked to name rather than rank the problems facing maize production. Qualitative and quantitative information on relevant aspects of these problems was collected and used to develop a more objective assessment of the relative importance of the various problems. Farmers' estimates of the following attributes were recorded, used to rank the production constraints mentioned by farmers, and used to identify appropriate strategies for research to deal with those problems:

- The percentage area affected by the problem.
- Average percentage damage caused (percentage yield loss).
- The season in which the problem occurs and its frequency of occurrence.
- The availability of a tolerant variety.
- The control measure used to deal with the problem.

The MDBP survey team chose this approach because there is a good chance that some of the factors listed here are not considered by farmers when they make their final evaluation of how serious a problem is. While it is important to consider how many farmers or hectares are affected by a problem (the first factor listed above), it is also necessary to assess how much damage or loss that problem causes (the second factor listed above). Some problems may affect fewer farmers or smaller areas but cause more serious damage (i.e., a 100% loss in yield). Some problems may cause serious damage in one season only, whereas others lead to moderate losses in the two maize crops grown every year. A problem that is very serious on all counts might occur only once every three to five years, unlike another problem that occurs yearly or in both seasons of the year.

An open-ended list of problems was developed so farmers could rank the elements and also add problems if they wished. The list was developed after extensive, country-wide informal surveys and field observations by multidisciplinary teams of maize researchers (Hassan and Karanja 1992). Alternative questions related to each of the attributes listed earlier (percentage damage caused by the problem, percentage area affected, and so forth) were tested at various locations in Kenya before the final wording of the questionnaire was identified (Hassan and Karanja 1992). Information on the first three attributes was used to calculate the percentage crop loss incurred because of a given problem each year. This figure represented farmers' estimate of the gross productivity gain expected from research directed at eliminating the problem. The gross efficiency index (G_i) defined in (5.1) was accordingly derived for each constraint. The following variables were employed as measures of the components of G_i:

- The expected proportional yield gain (Ey_i) was estimated by dividing farmers' estimate of the average proportional yield loss resulting from a problem by the frequency of its occurrence (in number of interval years). In other words, for a 30% yield loss caused by a problem that recurs every three years, a value of 0.10 (10%) expected yield gain per year was used.
- The gross productivity gain from eliminating problem i in season j ($Q_j \times Ey_i$) was multiplied by farmers' estimate of the proportion of maize area affected by problem i in season j (A_{ij}) to compute G_i in equation (5.3):

$$G_i = \beta \Sigma_j A_{ij} \times Q_j \times Ey_i, \tag{5.3}$$

where A_{ij} refers to the proportion of maize area affected by problem i in season j, Q_j is the baseline maize production realized in season j, and Ey_i is the expected yield gain from removing problem i (yield loss prevented).[2] Table 5.4 lists the seven most serious problems (those with the highest G_i) facing maize producers in zone i.

- The proportion of farmers using any measure to control or manage a problem was used as a proxy for the probability of success in dealing with the problem. In the case of problems for which no control measure is feasible, such as hailstorms and rain failure, the proportion of farmers indicating the availability of a tolerant variety was used as p_i.

The gross efficiency index (G_i) was derived to be the product of $p_i \times G_i$ (Table 5.5). The seven most serious problems identified in Table 5.4 were ranked on the basis of G_i from top priority (P1) to lowest priority (P7) within each zone (Table 5.5). The relative priority assigned to the problems varied significantly between zones. However, erratic or inadequate rainfall and stalk borers were important problems in all zones (Table 5.5). Hailstorms and stalk lodging caused the highest crop losses in the highest priority maize production zones (the highland tropics and moist transitional zone), whereas *Striga*, low soil fertility, and weeds other than *Striga* were the most serious constraints in the moist midaltitude zone, the third most important maize production zone. Weeds, other pests, and diseases were the biggest problems facing maize production in the other zones (Table 5.5).

The potential for breeding or crop management research to develop technologies and methods for reducing the problems identified by farmers was assessed by determining whether a genetic source of resistance/tolerance existed or whether farmers used a control measure. Farmers reported that their maize germplasm exhibited some tolerance of low soil fertility, stalk

Table 5.4. Farmers' estimates of yield losses and gross productivity gain (G_i) by type of problem and agroclimatic zone, Kenya.

	Lowland tropics	Dry midaltitude zone	Moist midaltitude zone	Dry transitional zone	Moist transitional zone	Highland tropics
Total maize production (000 t)	53	162	232	76	1234	909
March–August (long rains) (000 t)	45	122	170	45	1170	893
October–February (short rains) (000 t)	8	40	62	31	64	16
Stress factor/production constraint (000 t and proportion yield loss)						
Hail	– –	– –	– –	– –	357.9 (0.29)	190.0 (0.25)
Striga	– –	– –	97.5 (0.47)	– –	– –	– –
Rains[a]	11.2 (0.39)	50.2 (0.56)	62.7 (0.39)	33.5 (0.45)	135.8 (0.20)	299.9 (0.40)
Stalk lodging	– –	– –	– –	7.6 (0.19)	209.8 (0.21)	118.2 (0.24)
Stalk borer	13.5 (0.33)	27.6 (0.29)	60.3 (0.29)	19.8 (0.33)	172.8 (0.25)	100.0 (0.37)
Low soil fertility	– –	43.8 (0.62)	92.8 (0.40)	– –	111.1 (0.49)	– –
Chafer grubs	18.0 (0.34)	58.3 (0.44)	65.0 (0.34)	16.0 (0.21)	– –	– –
Weeds other than *Striga*	18.0 (0.34)	58.3 (0.44)	65.0 (0.34)	16.0 (0.21)	– –	82.9 (0.19)
Head smut	– –	– –	– –	45.6 (0.10)	123.4 (0.20)	100.0 (0.49)
Game animals	20.2 (0.54)	– –	– –	– –	– –	– –
Other pests[b]	18.0 (0.34)	51.9 (0.32)	67.3 (0.42)	21.3 (0.28)	– –	– –
Streak virus	– –	6.7 (0.34)	76.6 (0.50)	– –	135.8 (0.46)	– –
Other diseases[c]	11.7 (0.24)	– –	– –	– –	– –	– –

Source: MDBP.

Note: figures in parentheses are farmers' estimates of the proportion yield loss caused by problem i (Ey_i). Production realized in season j (Q_j) is multiplied by the proportion of maize area affected by a given problem and the percentage yield loss in the respective season. Units/numbers not in parentheses represent gross productivity gain (G_i) from total areas under maize affected by the problem in question in the zone, in 000 t.

[a] Both inadequate and erratic or unfavorable distribution of rainfall, depending on the zone in question.

[b] Pests other than stalk borer, chafer grubs, and game animals. The most important among these were termites, cutworms, weevils, grain borers, rodents, and birds.

[c] Diseases other than head smut and streak, such as stalk and ear rots and leaf blights.

Table 5.5. Maize production problems ranked for each agroclimatic zone according to gross efficiency criteria, Kenya.

Problem	Lowland tropics	Dry midaltitude zone	Moist midaltitude zone	Dry transitional zone	Moist transitional zone	Highland tropics
Hail	– –	– –	– –	– –	68.0 (3)	106.9 (2)
Striga	– –	– –	47.8 (1)	– –	– –	– –
Rains[a]	4.7 (2)	25.1 (2)	25.1 (2)	14.4 (2)	72.0 (2)	132.0 (1)
Stalk lodging	– –	–	– –	3.8 (7)	94.4 (1)	79.2 (3)
Stalk borer	2.6 (5)	11.1 (5)	19.9 (5)	7.1 (5)	63.9 (4)	46.1 (5)
Low soil fertility	– –	8.8 (6)	24.1 (3)	– –	42.2 (5)	30.0 (6)
Chafer grubs	1.0 (7)	13.2 (4)	– –	7.5 (4)	– –	– –
Weeds other than *Striga*	14.4 (1)	39.1 (1)	24.1 (3)	9.6 (3)	– –	– –
Head smut	– –	– –	– –	16.9 (1)	28.4 (7)	29.0 (7)
Game animals	4.7 (2)	– –	– –	– –	– –	55.0 (4)
Other pests[b]	4.5 (4)	14.0 (3)	18.2 (6)	6.0 (6)	– –	– –
Streak virus	– –	1.7 (7)	16.1 (7)	– –	34.0 (6)	– –
Other diseases[c]	2.4 (6)	–	– –	– –	– –	– –

Source: MDBP

Note: figures in parentheses indicate rank (1 = most important). Numbers not in parentheses indicate gross efficiency index $G_i = P_i \times G_i$.

[a] Both inadequate and erratic or unfavorable distribution of rainfall, depending on the zone in question.

[b] Pests other than stalk borer, chafer grubs, and game animals. The most important among these were termites, cutworms, weevils, grain borers, rodents, and birds.

[c] Diseases other than head smut and streak, such as stalk and ear rots and leaf blights.

Table 5.6. Kenyan farmers' assessment of the major constraints to maize production and strategies for overcoming them (complete sample).

Problem	Level of genetic tolerance/ resistance available	Tolerant/resistant variety used	Availability/ efficacy of control measures	Measure used
Hail	Low	H614, H625	None	–
Striga	High	Nyamula (local)	High	Weeding, roguing, manuring
Rains[a]	High	Katumani Composite (escapes problem)	None	–
Stalk lodging	High	H614, H625	None	–
Stalk borer	Low	Nyamula (local), H614	Low	Insecticide
Low soil fertility	Low	Nyamula (local), H614	High	Apply fertilizer
Chafer grubs	Low	Coast Composite	Low	Insecticide
Weeds other than *Striga*	Low	Nyamula (local), Coast Composite	High	Weeding, roguing
Head smut	High	H614, Katumani Composite, Nyamula (local)	Low	Cut and burn
Game animals	None	–	Low	Fencing and other methods of control
Other pests[b]	Low	H614	Low	Insecticide
Streak	Low	Nyamula (local), H614	None	–
Other diseases[c]	Low	H614	None	–

Source: MDBP.

[a] Both inadequate and erratic or unfavorable distribution of rainfall, depending on the zone in question.

[b] Pests other than stalk borer, chafer grubs, and large animals. The most important among these were termites, cutworms, weevils, grain borers, rodents, and birds.

[c] Diseases other than head smut and streak, such as stalk and ear rots and leaf blights.

lodging, head smut, *Striga*, and rain failure (Table 5.6). Many farmers used measures to control weeds (*Striga* and other weeds), other pests and diseases (stalk borers, chafer grubs, head smut, and other), and low soil fertility (Table 5.6). On the other hand, no practice enabled farmers to avoid hailstorms, rain failure, or stalk lodging. Farmers also knew no way of controlling streak virus and other diseases. The data in Table 5.6 were used to identify a suitable strategy for the KARI maize program to deal with the constraints that farmers identified. However, as mentioned earlier, the loss caused by each problem varies by zone, so research strategies will also vary, depending on the number of zones affected (Table 5.7). In general, the maize research program at KARI needs to direct its research resources toward reducing losses from:

Table 5.7. Farmers' ranking of maize production constraints compared with the research emphasis given to those constraints across agroclimatic zones of Kenya.

	Moist transitional zone		Highland tropics		Moist midaltitude zone		Dry transitional zone		Dry midaltitude zone		Lowland tropics	
Rank	1		2		3		4		5		6	
Suggested share of research resources[a]	29%		24%		19%		14%		9%		5%	
	(Rank)	(% expts.)[b]	(Rank)	(% expts.)	(Rank)	(% expts.)	(Rank)	(% expts.)	(Rank)	(% expts.)	(Rank)	(% expts.)
Major constraint[c]												
Pests (M/B)	2	(15)	2	(28)	5	(7)	4	(14)	3	(18)	2	(12)
Diseases (B)	4	(7)	4	(13)	6	(10)	1	(16)	5	(10)	4	(8)
Weeds (M)	–	(17)	–	(1)	4	(14)	3	(3)	1	(8)	1	(18)
Striga (B/M)	–	–	–	–	1	(6)	–	–	–	–	–	–
Low fertility (M/B)	3	(16)	3	(11)	3	(29)	–	(22)	4	(16)	–	(13)
Weather, stalk lodging (B)	1	(24)	1	(25)	2	(14)	2	(18)	2	(23)	3	(16)
Other	–	(21)	–	(22)	–	(20)	–	(27)	–	(25)	–	(15)
Research strategy[d]												
Breeding research	Shorter plants; medium- to late-maturing materials; more resistant materials		Shorter plants; more resistant materials		Tolerance of *Striga*; more resistant materials		Drought tolerance; higher yielding materials; more resistant materials		Drought tolerance; higher yielding materials; more resistant materials		Materials that mature faster; higher yielding materials; materials with greater resistance	
	67%	(52%)	67%	(53%)	55%	(36%)	72%	(69%)	50%	(58%)	40%	(28%)
Crop management research	Pest control, soil fertility management		Pest control, soil fertility management		Pest control, soil fertility management		Pest and weed control; water management		Pest and weed control; water management		Pest and weed control; water management	
	33%	(48%)	33%	(47%)	45%	(64%)	28%	(31%)	50%	(40%)	60%	(70%)

Source: MDBP.

[a] Proportional to priority points where the zone with the highest rank (1) is assigned six points and the zone with lowest rank (6) is assigned one point.

[b] '% expts.' indicates the percentage of experiments conducted between 1980 and 1992.

[c] The research strategy that stands the best chance of success (based on the results of Table 5.6) is provided in parentheses (B = breeding research, M = crop management research).

[d] Proportional to the priority rank of the identified constraint and named research strategy. Numbers in parentheses indicate the percentage share of the total number of experiments conducted between 1980 and 1992.

- Field and storage pests. Stalk borers, chafer grubs, and other pests (including termites, cutworms, weevils, grain borers, game animals, rodents, and birds) should be targeted for research, especially in drier areas and the lowland tropics. Farmers reported that their maize cultivars possessed very low to no resistance to these pests. However, farmers used several measures to control field and storage pests, indicating that crop management research has greater potential than breeding research for dealing with these pest problems.[3]
- Diseases. Farmers reported that the most serious diseases were head smut and maize streak virus (streak was a problem in the moist and dry midaltitude zones and the moist transitional zone) (Table 5.5). They reported observing some tolerance to head smut in their germplasm, but low resistance to streak and other diseases (Table 5.7). The data in Table 5.6 indicate that there is greater potential for solving these problems through breeding research than through developing alternative crop management practices.
- Farmers reported that low soil fertility was a major limitation to the productivity of their maize crops, especially in the most important maize-producing agroclimates (the moist transitional, highland tropical, and moist midaltitude zones) (Table 5.5). The unimproved maize variety Nyamula in the moist midaltitude zone, Katumani Composite, and H614 reportedly showed some tolerance to low soil fertility. Many farmers also reported using fertilizer to counter low soil fertility (Table 5.6), which suggests that there is scope for both breeding and crop management research to deal with this problem.
- Reported problems of hailstorms and stalk lodging indicate the need for shorter maize plants and stronger stalks in the moist parts of the transitional and highland tropical zones. Breeding would seem to be the appropriate strategy for addressing these problems.
- *Striga* and other weeds. Although tolerance to *Striga* was observed in local maize (Nyamula) in the moist midaltitude zone, where *Striga* is the most serious problem, low tolerance of other weeds was reported. However, most farmers used measures to manage both *Striga* and nonparasitic weeds.
- Erratic rainfall patterns. Irregular and/or inadequate rainfall was the greatest problem faced by maize farmers. Farmers said they observed good tolerance to erratic rainfall in their germplasm, though they knew of no control measure (Table 5.6). This would seem to indicate that a breeding strategy may overcome the problem, but earlier studies have shown that crop management can also help reduce losses associated with irregular rainfall in the semiarid zones (Wafula 1989; Keating *et al.* 1992).

Table 5.7 compares farmers' ranking of constraints with the share of maize research (measured as the number of experiments) dedicated to resolving these problems over the past 12 years. Whereas allocations of maize research resources between zones were found to deviate significantly from the efficiency/equity rankings in Table 5.3, allocations within zones among subprograms or disciplines were fairly consistent with farmers' efficiency ranking, except in a few cases (Table 5.7). For instance, farmers assigned a low priority to weeds in the moist transitional zone, whereas considerable effort was expended on weed research in this zone. The reverse is true for the dry transitional and dry midaltitude zones, where research on weeds received a much smaller share of the resources than farmers' priorities would dictate (Table 5.7). Another example is the relatively low share of research on *Striga* in past experiments in the moist midaltitude zone compared with farmers' evaluation of *Striga* as the most serious constraint in this zone. Moreover, the percentage of maize research resources invested in studying soil fertility problems in the dry transitional and lowland tropical zones was high, given farmers' assessment of the relative importance of these problems (Table 5.7).

An attempt was also made to derive the percentage shares of resources allocated to crop management versus plant breeding research, based on the ranking of problems by zone, the importance of each zone, and the kind of strategy identified for dealing with each problem (Table 5.7). The allocation of KARI research resources (numbers of experiments) between management and breeding research over the past 12 years was consistent with farmers' demand in the dry transitional and dry midaltitude zones. One problem with using the number of experiments as an indicator of resource allocation stems from the fact that breeding activities are not normally referred to as 'experiments.' A single breeding trial, for instance, might include a very large number of crosses and often occupies a much larger plot than a crop management trial might occupy (e.g., a nursery may occupy 1 acre). Therefore the number of breeding experiments may represent an underestimation of the actual resources expended per experiment for breeding research compared with crop management research.

However, results of the farmer survey indicate that more breeding research is required in the priority zones (moist transitional, highland tropical, and moist midaltitude) than has been the case in the past (Table 5.7). This conclusion could result from the fact that the maize program never targeted the moist transitional zone as a distinct adaptation zone. Researchers never developed a cultivar that would enable farmers to produce two maize crops each year – an important objective because of the area's high population density – in this zone's wet, relatively cool climate, which demands germplasm that takes a relatively long time to mature. Farmers in the highland tropics cited hailstorms, erratic rainfall, and stalk lodging as the most serious constraints, and solutions to these problems are also more likely come from breeding than crop management research. The fact that most farmers in the moist midaltitude zone use local cultivars indicates that a suitable improved variety is also lacking for this zone (Chapter 4). Nevertheless, it is important to note that low soil fertility, believed by researchers to be the major constraint in these zones (Ministry of Agriculture 1988, Smaling 1993), ranked third among farmers' priorities (Table 5.3). In fact, farmers may have underestimated this constraint.

Conclusion

Survey data on farmers' practices were combined with trial records and aggregate area and population statistics to evaluate the relevance of recent maize research to farmers' needs. Efficiency and nonefficiency criteria were used to rank maize production zones in order of priority. The moist transitional zone ranked first, followed by the highland tropics and moist midaltitude zones. In contrast, over the past 12 years the maize research program has emphasized research for the highland tropics most strongly, directing more than 50% of all experiments to that zone. Although the ranking exercise described in this chapter indicated that the moist transitional zone is likely to experience the greatest potential gains from research, this zone has received considerably less attention from research than the highland tropics. The share of maize experiments done in the moist midaltitude zone has also been low (1.9%) compared with that zone's relative importance (third in priority). These results suggest that KARI needs to reallocate maize research resources to the transitional and midaltitude zones if farmers are to realize greater benefits from research.

The agroclimatic zones differed significantly, however, in the kinds of problems limiting maize production and the kinds of research required to overcome those problems. Two problems appeared important in all zones: inadequate and erratic rainfall and stalk borers. Hailstorms and stalk lodging caused the greatest crop losses in the moist transitional zone and the highland tropics, whereas low soil fertility, *Striga*, and other weeds were the main constraints encountered in the moist midaltitude zone. Weeds, other pests, and diseases were the biggest problems facing maize production in the other environments. This variation among zones is one indication of why it is so important to target research to each zone to realize potential gains from research.

Except for a few instances, the allocation of maize research resources within zones – measured as the number of experiments conducted by different research subprograms over the past 12 years – was fairly consistent with the needs that farmers perceived. Farmers' assessment of priority problems indicated that more breeding research is required in the three most important zones (the moist transitional and moist midaltitude zones, as well as the highland tropics).

References

Alston, J., Norton, G., and Pardey, P. (1995) *Science Under Scarcity: Principles and Practices for Agricultural Research Evaluation and Priority Setting*. Cornell University Press, Ithaca, New York.

Cantrell, R.P. (1989) Recent developments in the CIMMYT Maize Program. In: *Toward Insect Resistant Maize for the Third World: Proceedings of the International Symposium on Methodologies for Developing Host Plant Resistance to Maize Insects.* International Maize and Wheat Improvement Center (CIMMYT), Mexico City.

Hassan, R., and Karanja, D. (1992) *Development of Survey Questionnaires and Organization of the Field Work.* KARI/CIMMYT Maize Data Base Project Document No. 5. Kenya Agricultural Research Institute (KARI) and International Maize and Wheat Improvement Center (CIMMYT), Nairobi, Kenya.

Hassan, R., Karanja, D., Arias, A., and Njoroge, K. (1992) *A Standardized Format for Organizing On-Farm and Experimental Maize Research Data in Kenya.* KARI/CIMMYT Maize Data Base Project Document No. 2. Kenya Agricultural Research Institute (KARI) and International Maize and Wheat Improvement Center (CIMMYT), Nairobi, Kenya.

KARI (Kenya Agricultural Research Institute) (1992) *Improve and Sustain Maize Production through Adoption of Known Technologies.* Information Bulletin No. 7. KARI, Nairobi, Kenya.

Keating, B., Wafula, B., and Watiki, J. (1992) Development of a modelling capability for maize in semi-arid eastern Kenya. In: Probert M.E. (ed.) *A Search for Strategies for Sustainable Dryland Cropping in Semi-arid Eastern Kenya.* Australian Centre for International Agricultural Research (ACIAR) Proceedings No. 41. ACIAR, Canberra, Australia.

Mills, B., Mwangi, P., Munyi, V., and Wambuguh, L. (1994) An analysis of production and price trends for maize, millet, sorghum, and wheat. Kenya Agricultural Research Institute (KARI), Socioeconomics Division. Memo, Nairobi, Kenya.

Ministry of Agriculture (Kenya) (1988) *Fertilizer Use and Recommendation Project: Main Report.* National Agricultural Research Laboratories, Nairobi, Kenya.

Otichillo, W., and Sinange, R. (1991) *Long Rains Maize and Wheat Production in Kenya in 1990.* Technical Report No. 140. Department of Remote Sensing and Resource Surveys, Ministry of Planning and National Development, Nairobi, Kenya.

Smaling, E. (1993) An agroclimatic framework for integrated nutrient management with special reference to Kenya. PhD thesis, Agricultural University, Wageningen, The Netherlands.

Trainer, T. (1989) *Developed to Death: Rethinking Third World Development.* Merlin Press, London, UK.

Wafula, B. (1989) Prospects for improving maize productivity through response farming. In: Probert, M.E. (ed.) *A Search for Strategies for Sustainable Dryland Cropping in Semi-arid Eastern Kenya.* Australian Centre for International Agricultural Research (ACIAR) Proceedings No. 41. ACIAR, Canberra, Australia.

Wood, S., and Pardey, P. (1993) *Agroclimatic Dimensions of Evaluating and Prioritizing Research from a Regional Perspective: Latin America and the Caribbean.* Discussion Paper No. 5. International Service for National Agricultural Research (ISNAR), The Hague, The Netherlands.

Endnotes

1 Research records were compiled from the National Agricultural Research Center at Kitale; the National Dry Land Farming Research Center at Katumani; the Regional Research Centers at Embu, Kisii, Kakamega, and Mtwapa; and the National Agricultural Research Laboratories in Nairobi.

2 Depending on the season of occurrence, Q_j is computed as follows (Table 5.4): for the long rains, if the problem occurs only during the long rainy season; for the short rains, if the problem occurs only during the short rainy season; for the two seasons,

if farmers indicated that they planted maize in both seasons (i.e., Q_j becomes Q and total annual production = $\Sigma_j Q_j$).

3 It is important to view such a conclusion with extreme caution, however, because it is risky to assume that if farmers do not observe genetic variability, then breeding will not be useful. Resistance to stalk borers, for instance, has been developed with success at CIMMYT (Cantrell 1989).

6 Estimating Potential Benefits from Research and Setting Research Priorities for Maize in Kenya

BRADFORD MILLS, RASHID M. HASSAN,
AND PETERSON MWANGI

A model was developed to estimate potential benefits to consumers and producers from different kinds of maize research in different agroclimatic zones of Kenya over a 30-year planning horizon. The types of research included breeding research, crop management research, and research on policies that affect farmers' use of new maize technologies. The simulated changes in total maize surplus and its producer and consumer components varied by type of research and target zone. The national maize program can expect to generate total benefits of 52 billion 1995 Kenya shillings (KSh) through the development and adoption of new varieties. Changes in surplus would accrue roughly equally to producers and consumers. In contrast, crop management research would yield total changes in surplus of a little over KSh 9 billion. Benefits from germplasm development would be concentrated in the moist transitional zone and highland tropics, which are large zones where the potential for developing new maize germplasm is good. The benefits from crop management research would be higher than those estimated for germplasm development in the dry and moist midaltitude and dry transitional zones. Potential gains from policy research amounted to about 30% of the expected total benefits. The *ex ante* research benefits estimated in this study should be compared with current resource allocations by zone and research theme to refine the research agenda and garner support for maize research based on its potential contribution to national development objectives.

Gains in productivity obtained as a result of research are an important potential source of growth in the agricultural sector. However, funding for agricultural research, particularly in developing countries, continues to decline, and managers of national agricultural research programs confront the difficult task of allocating a shrinking research budget between ever-increasing research needs. Because the impact of alternative research activities and technologies is unknown *a priori*, decisions about how to allocate research resources are best guided by a structured *ex ante* evaluation. More rigorous methods have been developed for *ex ante* evaluation of research. The tool kit of techniques now ranges from ranking alternatives based on subjective scoring criteria to the formal application of economic principles using a combination of subjective assessments of the potential for generating and adopting technologies and quantitative economic data (Alston *et al.* 1995).

The Kenya Agricultural Research Institute used the scoring method in its most recent evaluation of priorities across all commodity and factor research programs (KARI 1991). This exercise underscored the importance of maize to Kenya's agriculture and food economy: maize ranked first in importance among cereal crops and fourth among all commodities. However, priorities within the maize research program itself were not addressed. A subsequent strategic plan for cereal crop research (KARI 1994) unfortunately made only limited use of quantitative information, such as the MDBP data, or structured priority setting methods.

This chapter provides an example of how quantitative information can be used in a structured priority setting exercise, in which research priorities are based on an assessment of where research has the greatest potential to benefit producers and consumers of maize in Kenya. First, the economic surplus methods used in setting priorities for agricultural research are described, along with procedures for eliciting expert opinion on the potential for developing and adopting new technologies within each target agroclimatic zone. Results of an analysis of maize surpluses and deficits by zone and the price differences between zones are incorporated into a model of research impact with spatially linked production zones. The model generates estimates of potential consumer and producer benefits that could result from different kinds of maize research in the next 30 years. The chapter concludes with a brief discussion of how results of priority setting exercises such as this might be used by the national research program to make actual resource allocation decisions.

Measuring the Economic Benefits of Research

The most commonly used measure of the economic benefits from agricultural research is the change in economic surplus (consumer and producer surplus). *Consumer surplus* is the surplus value of a good to its consumers (the area under the demand curve in Fig. 6.1) over the price paid for the good. *Producer surplus* is the surplus of the price received for a good by its suppliers over the variable costs of its production (the area over the supply curve in Fig. 6.1). The changes in consumer and producer surplus resulting from a parallel, downward shift in supply induced by research are shown in Fig. 6.1 for a single period and market.[1] The important point to understand is that research benefits are linked directly to the magnitude of the rightward shift in the supply curve. Two components drive this shift: technology generation and technology adoption.

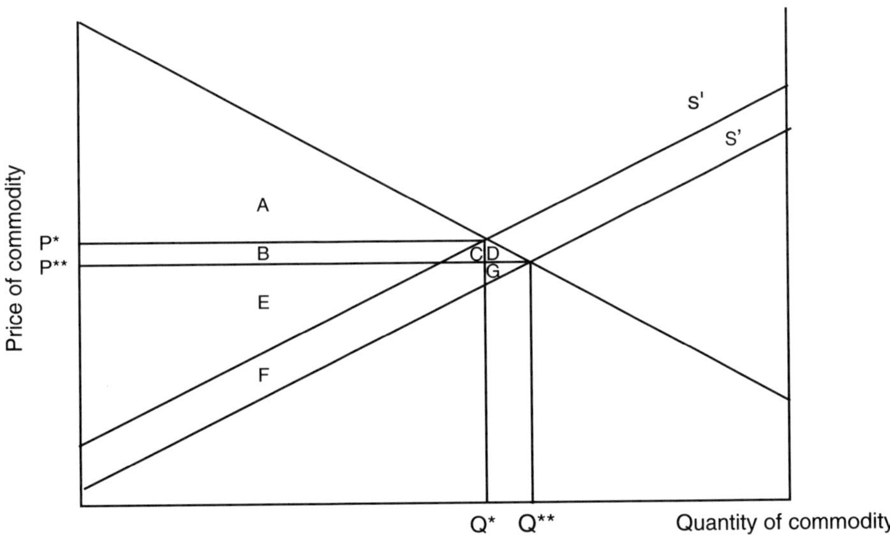

Fig. 6.1. Measuring the economic benefits from research.

Changes in consumer and producer surplus from a research-induced parallel supply shift from S to S′ are depicted here. On the pre-research supply curve, consumers receive area A in surplus. The commodity supply shift then results in a price decrease from P* to P** and consumers now receive economic surplus equal to the area A+B+C+D. The consumer benefit from research is given by area B+C+D.

The impact of the research-induced supply shift on producer surplus is more complex. Originally producer surplus is represented by the area E+B (the area below the commodity price but above the cost of production). After the supply shift, producer surplus is given by the area F+E+G. Thus the change in producer surplus is determined by two factors: gains from the unit cost reduction in production (area F+G) and losses from commodity supply shifts, through lower price (area B). Producer welfare may either increase or decrease as a result of technical change, depending on the relative size of the gains and losses. However, society as a whole (the aggregate of consumer and producer surplus changes) unambiguously gains (area C+D+F+G).

Technology generation

Because research is a process whose outcome is uncertain, the generation of technology through research is represented best as a distribution of possible outcomes. For research on a given agricultural commodity, outcomes are most commonly conceptualized in terms of yield increases obtained (or yield losses avoided). However, such yield increases often require additional inputs, which lower the effective value of yield gains. Therefore, research outcomes are specified as net yield gains, taking both yield increases and required additional input costs into account.

Potential minimum, most likely, and maximum net yield gains can be estimated using historical growth trends as a guide. A rigorous definition of what is commonly referred to as the 'probability of research success' can be used to account for research outcomes that farmers will never adopt.[2] Farmers,

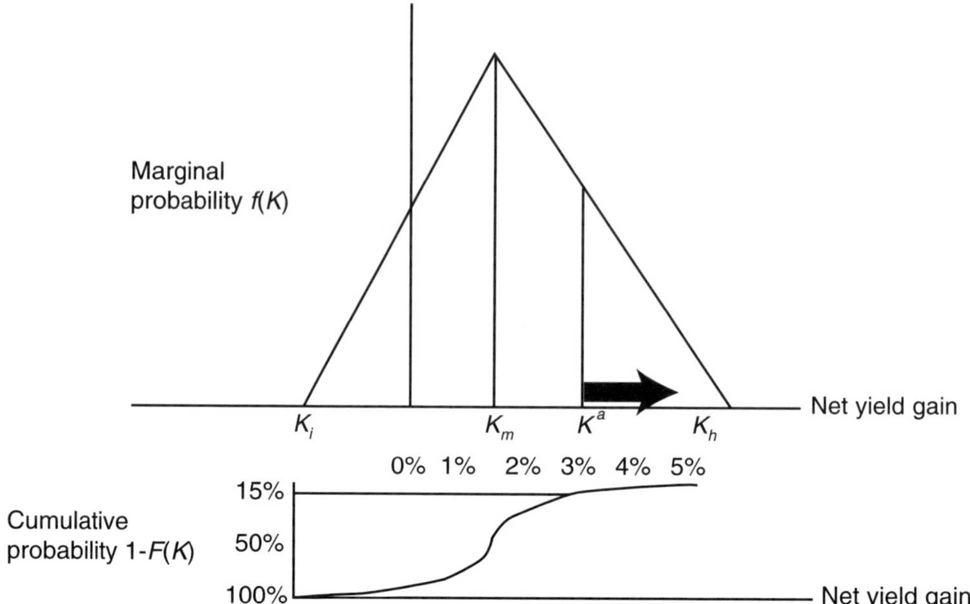

Fig. 6.2. Distribution of expected net yield gains.

Let K represent the net yield gain of an innovation. The minimum possible net yield gain is K_l; the most probable net yield gain is K_m; and the maximum net yield gain is K_h. The minimum net yield gain necessary for an innovation to be released for dissemination is K^a (a 3% gain in this example). For every K, there is a corresponding probability $f(K)$, which is assumed to follow a triangular probability distribution. The probability of achieving an increase in yields that is greater or equal to K^a, $\Pr(KK^a)$, is given by the cumulative density function $(1-F(k))$ above. In this example, the probability of a net yield gain from research above 3% is quite low, approximately 15%. The expected net yield gain is simply the expected value of K, conditional on K^a being achieved:

$$E\,[K \mid K \geq K^a].$$

particularly those who have few resources, will only adopt technologies if net yield increases are significantly greater than zero. This threshold level for adoption will depend on such factors as farmers' perceptions of the risk, additional labor investments, and additional capital investments associated with using the new technology. Technologies whose net yield gains do not exceed this threshold will not successfully pass through on-farm testing and evaluation and will not be released for dissemination. The threshold net yield gain necessary for technologies to be released should be defined for each major research theme and target agroclimatic zone.

For simplicity, the potential net yield increases achieved through research can be assumed to follow a triangular distribution (Fig. 6.2). The expected net yield gain is then calculated as the product of two parameters derived from the triangular probability density function: (i) the probability that a technology will exceed the net yield gain threshold and thus be released for dissemination and (ii) the expected net yield increase (conditional on the dissemination threshold being exceeded).

Technology adoption

Because the potential impact of research will depend on how rapidly and widely a yield-increasing technology is adopted, in calculating potential benefits from research it is essential to include an assessment of the probable adoption pattern. Figure 6.3 shows a generic adoption profile. Several basic characteristics of the profile are identified in the figure: (A) the research development lag, ending with the release of the new technology (at year 4 in the example); (B) the initially increasing adoption rate, which reflects the growing number of farmers in the target area who are using the technology; (C) the adoption plateau, occurring when most target farmers have been exposed to the technology and have decided whether or not to adopt it; and (D) declining adoption as the technology becomes obsolete. Together, these components determine the speed and frequency with which research results have an impact in farmer's fields. However, the profile parameters will also depend on the magnitude of the net yield gain embodied in the technology. Expected net yield gains, conditional on the dissemination threshold being exceeded, should be used as the basis for estimating potential adoption profiles. Again, historical adoption patterns can be a useful benchmark for estimating these parameters.

The full economic surplus model is presented in Appendix B. Two features of the model should be noted here. First, the shift in supply of a commodity attributable to research is simply calculated as the expected net yield gain multiplied by the period-specific adoption rate. The shift in supply for a given price can then be translated into a price (cost) change for a given supply through division by the supply elasticity.[3] Second, because the shift in supply

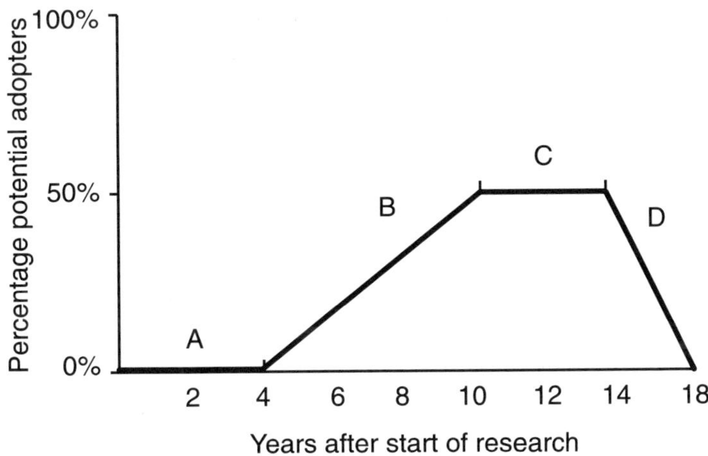

Fig. 6.3. Technology adoption profile.

is assumed to be parallel to the original supply curve, simple geometric formulas can be used to measure changes in consumer and producer surplus.

Data and Procedures

The basic objective of priority setting methods is to efficiently package information on potential research impacts to assist in resource allocation decisions. This section briefly explains how quantitative data from various sources and the subjective opinions of maize program scientists were used in setting priorities. A working group of key maize program scientists from different disciplines and regions of Kenya was formed to participate in this process.

Identification of research target zones and major research themes

Because the potential impacts of research will vary by agroclimatic zone, the MDBP zones (Plate 4) were used to identify the spatial domains within which the impact of particular agricultural technologies was expected to be relatively homogeneous. Production statistics for the six maize zones are given in Table 6.1.[4] The working group identified three major themes of the KARI maize research program to review in the priority setting exercise: varietal development research, crop management research, and technology environment research.

Varietal development research focuses on improving and/or maintaining the stability of maize yields through the manipulation of genetic material to develop new varieties. *Crop management research* focuses on improving agronomic practices for maize production. This theme is defined broadly to cover research on crop protection and soil fertility management. *Technology*

Table 6.1. Maize supply and demand by agroclimatic zone, Kenya.

Zone	Production (t)	Consumption (t)	Net surplus (t)
Lowland tropics	53,000	251,000	−198,000
Dry midaltitude zone	162,000	316,000	−154,000
Moist midaltitude zone	232,000	387,000	−155,000
Dry transitional zone	76,000	122,000	−46,000
Moist transitional zone	1,234,000	621,000	613,000
Highland tropics	909,000	427,000	482,000
Nonurban rest of Kenya[a]	0	311,000	−311,000
Nairobi	0	231,000	−231,000
Total	2,666,000	2,666,000	0

[a] Districts in regions where maize is not grown, primarily in the arid Northern Province.

environment research is relatively more complex. It may be defined in a narrow manner as research on improving the flow of existing technologies to extension agents and farmers. However, technology environment research may be defined more broadly to include research on ways to improve the policy and institutional environment within which technology transfer takes place. Whereas the activities of the Kenyan maize research program lie within the more narrow definition, socioeconomics research has a major role to play in the wider policy forum on improving the environment for the dissemination of research technologies.[5]

The potential for generation and adoption of technologies

As indicated earlier, the outcome of research investments will not be realized for many years, and *ex ante* parameters of technology generation and adoption must be based on the subjective opinion of informed sources. Because the most knowledgeable sources for this kind of information are experts in the research program, who have a vested interest in the outcomes of the priority setting exercise, an attempt must be made to control for bias. The reliability of experts' estimates is often further compromised by a poorly developed conceptual framework for eliciting the parameters of technology generation and adoption.

The KARI working group tried to eliminate potential bias in three ways. First, a facilitator collected benchmark information on historical yield and production growth trends in the target zones, along with information from the MDBP on the adoption of technologies released by the research program. This information served as a reference point when the group assessed the potential for technology generation and adoption. Second, the group setting provided some control over bias as the working group reviewed the major constraints to maize production in each zone, identified the key problems to be addressed by the three research themes, and identified the research impacts that could be made by addressing these constraints. If an individual scientist happened to promote a particular area of research too strongly, the rest of the group, with other vested interests, could challenge the information being provided and develop a consensus based on revised assumptions. Finally, the working group was expected to present its assumptions on the potential for generating and adopting technology to a larger group of persons concerned with the direction and outcomes of maize research, including farmers and input and processing industry representatives, for review and in some cases modification. This process of client review placed further pressure on the group to provide realistic estimates. The assumptions that guided the working group are presented in Appendix C.

Market information

The structure of maize markets is an important determinant of the magnitude of research benefits. Information on net maize balances, relative prices, and the exogenous demand shifts expected to result from population growth in each target agroclimatic zone was collected from publications of the Ministry of Agriculture and CBS. Demand and supply elasticities relevant to maize, which affect both the absolute and relative magnitude of research benefits to producers and consumers, were available from previous studies. The next section summarizes results of the analysis of maize markets in Kenya, which are used later in estimating the potential benefits of maize research.

Characteristics of Maize Markets in Kenya

In an average year, Kenya produces just enough maize to meet internal demand, and official maize imports and exports are limited. Statistics from FAO indicate that between 1989 and 1992 net imports of maize in Kenya averaged 24,000 t, equivalent to less than 1% of production (FAO 1994). In three of those four years, Kenya recorded no official international trade in maize.[6] Nyoro (1992) suggests that these low levels of international trade result from high transaction costs in import and export markets, which effectively close Kenyan maize markets.[7]

Based on the low level of reported external trade, the aggregate demand for maize within Kenya was assumed to be equal to supply when current net surpluses were calculated for each zone. The distribution of maize consumption across districts was then calculated based on household maize consumption estimates from the 1979 Kenya Rural Household Budget Survey and 1989 district-level census records on the number of households per district (CBS 1982, 1994).[8] Finally, the consumption of maize was allocated to target zones based on the proportion of each district within each target zone. The resulting figures were compared with production estimates for the target zones (Table 6.1) to derive estimates of current net maize surplus by zone. The results show that all zones except the moist transitional zone and highland tropics are net importers in an average year. Nairobi, nonurban areas in the rest of Kenya, and deficit zones rely on the surpluses produced in the high-potential zones to meet their consumption needs through trade.

Price differences from zone to zone were estimated empirically based on 1992–1993 monthly retail prices of maize per kilogram for markets across Kenya (supplied by CBS) and on 1992–1994 retail prices for 90-kg bags of maize in major markets (supplied by the Ministry of Agriculture's Market Information Branch) (Mills *et al.* 1994). Markets were spatially referenced and Thiessen polygons were constructed to allocate all areas within Kenya to the nearest market (Eastman 1992). Mean monthly maize prices for each zone, weighted by area, were then calculated along with monthly prices for the capital city of Nairobi.[9] The estimated mean price in each zone and the price difference relative to Nairobi are given in Table 6.2 for a tonne of maize in December 1994.

The highest price for a tonne of maize was found in the lowland tropics, reflecting the high cost of transporting maize to this area from surplus regions in western Kenya and the correspondingly high import parity price. The price

Table 6.2. Real maize prices (KSh) by zone, Kenya, 1992–1994 (at December 1994 prices).

Zone	Mean	Price difference between zone and Nairobi
Lowland tropics[b,c,d,e,f,g]	15,070	2030
Dry midaltitude zone[a,c,d,e]	13,400	360
Moist midaltitude zone[a,b,d,e,f,g]	12,180	−860
Dry transitional zone[a,b,c,g]	12,590	−550
Moist transitional zone[a,b,c,f]	12,540	−500
Highland tropics[a,c,e]	12,930	−110
Nairobi[a,c,d]	13,040	0

Note: Paired *t*-tests show the mean to be statistically significantly different at the 5% level from (a) the lowland tropics, (b) the dry midaltitude zone, (c) the moist midaltitude zone, (d) the dry transitional zone, (e) the moist transitional zone, (f) the highland tropics, and (g) Nairobi.

of maize in the dry midaltitude zone was also higher than in Nairobi, although the difference was not statistically significant. By contrast, the moist midaltitude, dry transitional, moist transitional, and highland tropical zones all showed lower prices than Nairobi, with the difference being statistically significant in the moist midaltitude and dry transitional zones. These price differences are assumed to reflect the real transport costs of moving maize between surplus zones and Nairobi. Thus, in the spatially linked production zones model, cross-zone price differences are held constant in absolute terms as prices rise and fall because of research-induced shifts in supply and other exogenous influences on maize supply and demand in specific zones.

While research is the source of changes in supply, the most important factor influencing demand for maize in Kenya is population growth. Population growth rates for each zone were calculated from 1979 and 1989 district census estimates, again assuming that population is proportionally distributed by the area of each zone in each district (CBS 1994). Table 6.3 shows that between 1979 and 1989 the Nairobi area had the highest rate of population growth, 4.7% per annum, while the nonurban areas in the rest of Kenya (primarily the Northern Province Districts) showed the lowest rate of growth at 1.2% per annum. The six maize production zones all showed very high rates of population growth, ranging from 3.0% to 3.9% per annum. These growth rates were discounted by 25% in the model to reflect projected decreasing rates of population growth. On the supply side it is assumed that there are no exogenous sources of growth other than the research-induced shifts of supply. This assumption is reasonable, given that little land suitable for maize can be brought into production and that maize area under production has remained fairly constant over the past 20 years.

Table 6.3. Past and projected rates of population growth by maize agroclimatic zone, Kenya.

Zone	Population growth 1979–1989 (%)	Projected (−25%)
Lowland tropics	3.06	2.29
Dry midaltitude zone	3.40	2.55
Moist midaltitude zone	3.07	2.30
Dry transitional zone	3.89	2.92
Moist transitional zone	3.25	2.44
Highland tropics	3.69	2.77
Nonurban rest of Kenya[a]	1.24	0.93
Nairobi	4.70	3.52
Aggregate	3.36	2.52

Source: CBS (1982, 1994).

[a] Districts in regions where maize is not grown, primarily in the arid Northern Province.

Finally, the shape of the shifts in the supply and demand curves will have an important impact on the magnitude, and particularly the distribution between producers and consumers, of estimates of economic surplus resulting from research. In the absence of contrary information, supply and demand curves are assumed to be linear and to shift in a parallel fashion. The actual slopes of the curves are determined by the supply and demand elasticities for maize in Kenya. Kiori and Gitu (1991) estimated a long-run supply elasticity of 0.684 for Kenya. This estimate is close to the center (mean) of the distribution of long-run supply elasticities for commodities in developing countries and is used in the current study.[10]

Estimates of the elasticity of demand specifically for maize are also available from several studies in Kenya. Bezuneh *et al.* (1988) estimated the own-price demand elasticity for maize and beans to be −1.19 in an almost ideal demand system for rural households in Baringo District. Jayne *et al.* (1995) calculated urban own-price elasticities of −1.41 and −0.11 for sifted and whole maize flour, respectively, in urban Nairobi. Finally, Dorosh *et al.* (1994) estimated an own-price elasticity of −0.826 in an almost ideal demand system for a consumption system in Mozambique that, as in Kenya, relies on white maize. Based on these results, a demand elasticity of −1.0 was considered a reasonable estimate for maize for all zones in the study.

Research Benefits in Spatially Linked Production Zones

A multiperiod, multiple interlinked market version of the basic model of economic surplus described earlier was used to estimate the benefits from maize research in Kenya. The full model specification, given in Appendix B, follows the multimarket model described in Appendix 5.1.2 of Alston *et al.* (1995). The model accounts for the dynamic elements of agricultural research measured in the previous sections, including the specific time profiles for technology generation and adoption; increased demand for commodities as a result of population growth; variable prices across production zones; and the impact of research-induced price spillovers to other production zones.

Results

The model was solved recursively using an EXCEL spreadsheet. Table 6.4 presents the estimated research-induced changes in producer and consumer surplus. These results represent the aggregate of the direct research impacts for each zone and the indirect impact through research-induced price changes in other maize production and consumption zones. A real discount rate of 5% is used.

The simulated changes in total surplus and its producer and consumer components vary markedly across research themes and target zones.[11] The KARI maize research program can expect to generate total benefits of 52 billion 1995 Kenya shillings (KSh) through research to develop new varieties and hybrids. Changes in surplus would accrue roughly equally to producers (KSh 31 billion) and consumers (KSh 21 billion). In contrast, crop manage-

Table 6.4. Net potential economic surplus (million KSh, real discount rate = 5%) generated by potential KARI maize program research, by type of research and across agroclimatic zones, Kenya.

Type of research/ surplus	Lowland tropics	Dry mid-altitude	Moist mid-altitude	Dry transitional	Moist transitional	Highland tropics	All zones
Varietal development							
Producer surplus	97.5	0	400.0	808.7	27,435.9	2,609.7	30,968.7
Consumer surplus	53.6	0	292.1	531.4	17,988.1	1,751.5	20,758.8
Total surplus	151.1	0	692.0	1,340.2	45,424.0	4,361.2	51,727.5
Crop management							
Producer surplus	110.8	1,448.5	1,166.3	912.3	1,610.2	159.5	5,730.2
Consumer surplus	60.7	893.6	840.2	602.8	1,142.3	108.8	3,663.0
Total surplus	171.5	2,342.1	2,006.5	1,515.1	2,758.9	268.3	9,393.2
Technology environment							
Producer surplus	59.6	490.0	934.1	265.6	2,534.7	2,974.9	7,185.4
Consumer surplus	33.0	311.4	669.4	182.7	1,785.2	1,992.5	4,999.0
Total surplus	92.5	801.4	1,603.5	447.9	4,328.8	4,967.4	12,184.4

ment research would yield total changes in surplus of a little over KSh 9 billion.

The benefits from varietal development are highly concentrated in the moist transitional zone, mainly because this zone encompasses a large area and there is good potential for developing a maize variety that matures more rapidly than the 600-series hybrids and more slowly than the 500-series. Substantial benefits are also expected from varietal development research in the highland tropics – again, primarily because of the size of the zone (the working group actually felt that the potential for generating net yield gains through varietal development in this zone was rather limited – see Appendix C). Only moderate benefits are expected from developing varieties for the dry transitional and moist midaltitude zones, and minimal benefits for the lowland tropical and dry midaltitude zones.

However, within several other zones the benefits from crop management research are higher than those estimated for varietal development. The smaller production zones (dry and moist midaltitude and dry transitional zones) show great potential for generating economic gains from crop management research because of the large expected net yield increases (Table 6.4). The high potential for crop management in the moist transitional zone can be attributed primarily to the large production base. In fact, the probability of developing technologies suitable for dissemination in the moist transitional zone scored very low in the working group's assessment (Appendix C). Finally, estimated surplus benefits are quite small in the lowland tropics (owing to the small production base) and highland tropics (owing to the low probability of developing technologies with the minimum net yield gains necessary for successful dissemination).

The potential gains from the technology environment theme are significant, amounting to about 30% of the expected total benefits (Table 6.4). These benefits are directly linked to the production base in each zone, because the working group's predictions do not show large variations between zones in the potential for generating and adopting results of technology environment research. Owing to their size, the moist transitional and highland tropical zones show the highest estimated benefits, despite recording the lowest expected net yield increases from this type of research. On the other hand, the dry midaltitude, dry transitional, and lowland tropical zones show little potential for generating benefits through technology environment research because of their small production bases.

In all cases, producer surplus changes comprise a slightly higher proportion of total surplus changes than consumer surplus. However, when the consumer and producer benefits occurring only within the target zone for research (Table 6.5) are compared with net benefits across all zones (Table 6.4), consumer benefits from research in a specific zone are found to be widely distributed across all zones. Changes in producer surplus, in contrast, are highly concentrated within the target zone for research and are negative in other zones. These results stem from the fact that under the assumption of interlinked markets, real maize prices will be lower in all zones under the 'with-research' scenario compared with the 'without-research' scenario.

Table 6.5. Potential economic surplus (million KSh, real discount rate = 5%) generated within each agroclimatic zone of Kenya by KARI maize research.

Type of research/ surplus	Lowland tropics	Dry midaltitude	Moist midaltitude	Dry transitional	Moist transitional	Highland tropics
Varietal development						
Producer surplus	150.1	0	666.2	1,324.3	36,664.4	3,760.4
Consumer surplus	5.1	0	40.0	26.9	4,131.8	303.2
Total surplus	155.2	0	706.2	1,351.2	40,796.2	4,063.6
Crop management						
Producer surplus	170.4	2,287.4	1,932.0	1,395.4	2,226.6	231.4
Consumer surplus	5.5	109.9	116.1	30.2	262.6	18.9
Total surplus	175.9	2,397.3	2,048.1	1,425.6	2,489.2	250.3
Technology environment						
Producer surplus	91.9	782.4	1,543.6	442.7	3,494.0	4,283.1
Consumer surplus	3.1	38.3	91.3	9.4	407.0	344.8
Total surplus	95.0	820.7	1,635.0	452.1	3,901.0	4,627.9

Producers in the targeted zone are compensated by the additional surplus generated from lower unit costs, which outweigh losses from lower prices (Fig. 6.1). However, producers in other nontarget zones face lower equilibrium prices without accompanying unit cost reductions and thus show negative changes in producer surplus. Consumers, in contrast, enjoy surplus increases from lower prices in all zones. Therefore, changes in consumer surplus from zone-specific research are larger when direct and indirect impacts are aggregated across all zones than when consumer surplus changes only within the target zone are considered.

The equilibrium dynamics of the model also suggest some sobering future trends for maize markets in Kenya. Even under the most optimistic research scenarios, given the extremely high rate of population growth in Kenya, supply will not keep pace with demand for maize and real prices are predicted to almost double over the next 30 years. Under a more realistic scenario, real maize prices rise to a level where formal imports become economically feasible, and world market prices plus transaction costs become the basis for setting internal market prices in Kenya. However, under this scenario, Kenya will need to import a significant proportion of its major staple.

Conclusion: Linking Model Results to Resource Allocation Decisions

The goal of the exercise described in this chapter was to assist the KARI maize program in setting research priorities across agroclimatic zones and research activities. This kind of structured priority setting provides scientists in the maize program with a better understanding of the environment for technology development. The classification of agroclimatic zones helps identify areas associated with key biophysical parameters for maize production. The zoning scheme also allows maize researchers to examine potential spillovers to other research centers in the region. A review of the current production situation within zones, based on historical data sources, improves researchers' understanding of the trends related to maize production and consumption that must be addressed. Finally, a systematic analysis of the potential for generating research results and adopting research products in the current production environment improves the research program's understanding of potential impact.

The *ex ante* research benefits estimated in this study should be compared with current resource allocations (particularly human resources) within the KARI maize program by zone and research theme. The results can help KARI to refine its research agenda and make a coherent argument about its potential contribution to national agricultural development objectives. However, priority setting in the final analysis is an institutional process for making resource allocation decisions. The usefulness of a priority setting exercise such as the one described here can be improved dramatically if it is integrated with an institutional process for establishing resource allocation guidelines for the research program (Stewart 1995). To achieve this objective, the information generated by the working group will be presented to a larger group of stakeholders in the maize program's research, as described earlier. The stakeholder group will be asked to examine the assumptions underlying the results described here and modify them where necessary. The group will also compare these potential benefits with current human and financial resource allocations within the KARI maize program. The results of this analysis will become the basis for defining major maize research themes, by zone, as high, medium, and low priorities for the research program. Specific guidelines will be set to indicate how research managers should adjust allocations over the medium-term planning horizon.

References

Alston, J., Norton, G., and Pardey, P. (1995) *Science Under Scarcity: Principles and Practices for Agricultural Research Evaluation and Priority Setting*. Cornell University Press, Ithaca, New York.

Bezuneh, M., Deaton, B., and Norton, G. (1988) Food aid impacts in rural Kenya. *American Journal of Agricultural Economics* 70(1), 181-191.

CBS (Central Bureau of Statistics) (1982) *Rural Household Budget Survey: Basic Report*. Government Printer, Nairobi, Kenya.

CBS (Central Bureau of Statistics) (1994) *Kenya Population Census, 1989*. Government Printer, Nairobi, Kenya.

Dorosh, P., delNinno, C., and Sahn, D. (1994) *Food Aid and Poverty Alleviation in Mozambique: The Potential for Self-targeting of Yellow Maize*. Cornell Food and Nutrition Policy Program, Ithaca, New York.

Eastman, J.R. (1992) *IDRISI, Version 4: Technical Reference Manual*. Clark University, Graduate School of Geography, Worcester, Massachussetts.

FAO (Food and Agriculture Organization) (1994) *AGROSTAT PC*. FAO, Rome, Italy.

Jayne, T., Lupi, F., and Mukumbu, M. (1995) *Effects of Food Subsidy Elimination in Kenya: An Analysis Using Revealed and Stated Preference Data.* Department of Agricultural Economics Staff Paper No. 95–23. Michigan State University, East Lansing, Michigan.

KARI (Kenya Agricultural Research Institute) (1991) *Kenya Agricultural Research Priorities to the Year 2000.* KARI, Nairobi, Kenya.

KARI (Kenya Agricultural Research Institute) (1994) *Strategic Plan for Cereals in Kenya (1993–2013).* KARI, Nairobi, Kenya.

Kiori, G., and Gitu, K. (1991). *Supply Response in Kenya Agriculture: A Disaggregated Approach.* Ministry of Planning and National Development Technical Paper 92–11. Ministry of Planning and National Development, Nairobi, Kenya.

Mills, B., Mwangi, P., Munyi, V., and Wambuguh, L. (1994) An analysis of production and price trends for maize, millet, sorghum, and wheat. Kenya Agricultural Research Institute, Socio-Economics Division. Memo, Nairobi, Kenya.

Nyoro, J. (1992). Competitiveness of maize production systems in Kenya. In: *Proceedings of the Conference on Maize Supply and Marketing under Market Liberalization, Institute of Banking and Finance, 18–19 June, Karen-Nairobi, Kenya.* Egerton University, Njoro, Kenya. pp. 10–39.

Rao, J. (1989). Agricultural supply response: A survey. *Agricultural Economics* 3, 1–22.

Stewart, J. 1995. Models of priority setting for public research. *Research Policy* 24,115–126.

Endnotes

1 See Chapter 2 in Alston *et al.* (1995) for a more complete discussion of economic surplus measures in agricultural research evaluation.

2 In the standard framework for calculating research benefits, probability of success is multiplied by the expected net yield gains across the full distribution of possible outcomes. This practice erroneously includes research outcomes that will never be adopted in farmers' fields in the calculation of expected net yield gains.

3 Supply elasticity is a commonly used economic measure of a commodity's responsiveness to price. Measures of elasticity for maize in Kenya are discussed later in this chapter.

4 Production figures used here correspond to those given in Chapter 5.

5 It can be argued that efforts to disseminate technology translate into research benefits in a different way than do other kinds of research. However, in analyzing the potential of technology environment research to generate benefits, the working group used the same procedures used in analyzing the potential of other kinds of research. To minimize the overlap between benefits from technology environment research and other kinds of research, the working group focused on the potential for increasing the flow of currently available technologies to farmers' fields.

6 These figures do not include cross-border trade, particularly with Uganda in the west. There are no quantitative estimates of the net magnitude of this trade or its impact on the net surplus of maize in western Kenya.

7 Based on April 1992 prices, Nyoro (1992) calculated a Nairobi import parity price of KSh 614 per 90 kg bag and an export parity price of KSh 195 per bag. Domestic producer prices ranged from KSh 300 to KSh 400 during the same period.

8 No figures for maize consumption at the household level exist for the urban centers of Nairobi and Mombasa or for districts in the North Eastern Province. National

average household levels were assumed to represent the pattern of consumption in these areas.

9 The number of markets was insufficient to calculate a monthly zonal average for the regions outside Nairobi where no maize is grown.

10 Rao (1989), in a review of empirical supply elasticity estimates, found that long-run estimates generally ranged between 0.3 and 1.2.

11 It is important to note that KARI maize scientists' estimates of the potential for generating and adopting technologies are the basis of the results presented here. For example, the fact that these researchers estimate a zero probability of developing varieties with sufficient net yield gains for dissemination in the dry midaltitude zone (Appendix C, pp. 210–220) implies that there will be zero gains in consumer and producer surplus from research to develop new maize varieties for this zone.

III Patterns of Maize Technology Diffusion and the Impact of Research

7 Adoption Patterns and Performance of Improved Maize in Kenya

Rashid M. Hassan, Kiarie Njoroge, Mugo Njore, Robin Otsyula, and A. Laboso

Patterns of adoption of improved maize were analyzed by agroecological zone and farmer group, using a qualitative response model (the logit model) to test hypotheses about several determinants of adoption. The adoption of modern maize materials was greater and more rapid in high-potential zones compared with the semiarid zone and lowland tropics. Farmers also adopted hybrid maize more rapidly than improved open-pollinated varieties (OPVs). Larger gains in yield and better postharvest qualities are needed for improved OPVs to replace local cultivars in the low-potential areas. Investment in improving seed distribution, extension, and the physical infrastructure in low-potential areas is crucial for wider adoption of improved maize seed technology; seed pricing policies also need to reflect the differential gains from different kinds of improved seed. Increased use of fertilizers in the high-potential areas could deliver significant gains in maize productivity. Although there is a demand for research to develop hybrids and crop management practices that farmers can use in areas where soil fertility is poor, pricing and marketing policies can also play an important role in addressing the limitations of poor soil fertility by promoting the use of commercial fertilizers. Better education and credit facilities may also lead to increased productivity among smallholders through higher use of commercial fertilizers. Policies and institutional arrangements that would improve the access of younger farmers to land could also promote wider adoption of improved maize.

The rapid adoption of hybrid maize in Kenya is often cited as one of the success stories of hybrid seed technology in less developed countries (Gerhart 1975; USAID 1980; Blackie 1989; Byerlee 1994a; Sarch and Gilbert 1995). The first hybrid maize variety was released in 1964; ten years later, all of the large-scale farmers and a significant proportion of the smallholders in the medium- to high-potential zones of Kenya were planting hybrids (Gerhart 1975). In addition to the large yield advantage that these hybrids had over farmers' local materials, the establishment of an organized seed multiplication and distribution system and a relatively good infrastructure were crucial factors in the adoption of hybrid maize in these zones. However, adoption of hybrids had been much slower among farmers in the low-potential zones of western Kenya.

In the years since Gerhart's study was done, no comprehensive investigation of the adoption of improved maize – hybrids as well as open-pollinated varieties (OPVs) – has been conducted in Kenya, apart from a few studies over limited areas (e.g., Ongaro 1991; Mohamed *et al.* 1985). This chapter examines the extent to which improved maize has continued to spread among farmers in Kenya's high-potential areas and the extent to which it is moving into less favorable zones.

Our study expands on earlier work in several important ways. First, because Kenya's maize breeding program has continued to release new varieties over the years, the study seeks to assess the impact of new cultivars. A measure of the rate at which farmers replace older cultivars with newer releases is derived to examine the efficiency of varietal turnover and diffusion. Second, some important maize-producing regions of Kenya were not covered by Gerhart's survey, including the midaltitude zones and lowland tropics. Although some hybrids developed for the highlands have been used in these zones, hybrids and OPVs have also been developed and released especially for these areas (Harrison 1970; Njoroge *et al.* 1992b, 1992c, 1992d). Third, our study departs from earlier studies by comparing the suitability, performance, and adoption of improved maize across agroclimatic zones and groups of farmers. Farmers' preferences for varietal traits other than yield are examined, and the factors affecting adoption of improved seed technology are analyzed.

The ultimate objective of this work is to determine the different stages of agricultural transformation, hence the different technological needs and potential sources of gains in maize productivity, in Kenya's diverse agroclimatic zones and farmer groups. This information will facilitate evaluation of maize improvement research, extension, and seed supply services. In addition, improved knowledge of the status and patterns of seed technology adoption will help maize researchers choose appropriate breeding strategies for developing the array of technologies that will be needed by different groups of farmers in different agroclimatic zones.

Data Sources

This study is based mostly on information from the geo-referenced survey of Kenyan maize farmers described in Chapter 3. During the survey, farmers were asked about their history of seed technology adoption, including information on when they first began planting improved maize. Information on current practices, such as the type of seed used by the farmers, area sown to maize by cultivar, and sources of maize seed and information about maize technology, was recorded together with information on other production practices and cropping systems information. At the village level, data were collected on input supply, extension, and other marketing infrastructure.

Maize Improvement Research in Kenya

Formal maize improvement research began in earnest in Kenya in 1955, when a full-time breeder was posted to Kitale Research Center (Gerhart 1975). The Kitale program, whose mandate is to develop late-maturing materials suited to the wet conditions of the highland tropics, has released 14 hybrids since 1961. These hybrids came from two sources: Kitale Synthetic II (KSII) and Ecuador 573 (EC573) (Table 7.1). The first source, KSII, is an OPV selected from farmers' local material, known as 'Kenya Flat White Maize,' and released in 1961 (Appendix D). Kenya Flat White Maize is a mixture of landraces imported from the US, South Africa, and the West Indies and adapted by European settlers (Harrison 1970). The second source of the Kitale hybrids, EC573, is unimproved local stock imported from Central America in 1959 to cross with KSII and enhance the genetic diversity of Kenyan hybrids (Harrison 1970). These two populations were continuously improved at Kitale through reciprocal recurrent selection, and they still constitute the genetic base of all Kitale hybrids (Njoroge *et al.* 1992a). However, two broadly based composites (Kitale Composites B and E) have been developed to expand the narrow genetic bases of KSII and EC573 and serve as long-term back-up populations for reciprocal recurrent selection at Kitale (Ogada 1975).

The second oldest maize breeding program in Kenya was established at Katumani in 1957. Its objective is to develop improved varieties suitable for the semiarid conditions of eastern Kenya (Dowker 1964). Over the past 30 years, three composites have been developed at Katumani, based on two synthetic maize series (Table 7.1).[1] Breeders selected the first series, the Katumani Synthetics, from 'Machakos Local White,' a local landrace that is a variant of Kenya Flat White Maize (F. Ogada, unpublished paper). Breeders developed the second series from combinations of the Mexican unimproved variety, Taboran ('Taveron') and other introductions, such as some of the first US hybrids, hybrids from Israel, and a French composite (Harrison 1970). The Katumani program has continued to improve and recombine these two populations to achieve further genetic gains. Unlike the breeding program in Kitale, the Katumani program has continued to dedicate its resources to the development of composites rather than hybrids.

Between 1957 and 1965, two other maize breeding programs were added to the national public research system. The midaltitude maize breeding program at Embu was established in 1965 to develop improved medium-maturing varieties and drought-escaping materials suitable for areas in central Kenya with a bimodal rainfall pattern (Njoroge *et al.* 1992c). The Embu program has released two hybrids, H511 and H512, developed from the composites E11 and E12 (improved late and early varieties from the Kitale and Katumani programs) and the variety SR52 (a single-cross hybrid used throughout southern Africa) (Njoroge *et al.* 1992c). The fourth maize breeding program is based at Mtwapa Research Center. The Mtwapa program develops improved cultivars to fit the hot, humid environment of the lowland tropics along the coast. Incorporating disease resistance into maize, especially resistance to lowland rust, has been a major objective (Njoroge *et al.* 1992c) of this program. A

Table 7.1. Maize improvement research programs, mandate areas, number of maize materials released, and their genetic base, Kenya.

Program and year established	Type of germplasm required (days to maturity)	Elevation range (masl)	Rainfall range (mm)	Number of materials released	Name and and year of recent release	Genetic source of release
Kitale (1955)	Late maturing (> 180 days)	>1800	>1000	14	H626 (1989)	a) Kitale Synthetic II b) Ecuador 573
Katumani (1957)	Early maturing (90–120 days)	500–1600	400–800	3	Makueni Composite (1989)	a) Katumani Synthetics b) Taboran (Taveron) c) Other[a]
Embu (1965)	Medium maturing (120–150 days)	1500–1800	700–1800	2	H512 (1970)	a) Embu E1, E2, E11, E12 b) SR52
Mtwapa (1957)	Lowland tropics (90–120 days)	0–500	500–1000	2	Hybrid Pwani (1989)	64 local and exotic lines[b]

Source: Harrison (1970), MDBP survey of experimental records, 1992; Njoroge (1992a, 1992b, 1992c, and 1992d).

[a] Local yellow flints, early US hybrids, Israeli hybrids, and a French composite (Harrison 1970).

[b] Details in Njoroge *et al.* (1992d).

broad-based composite (Coast Composite) was developed in 1974 from a wide range of local and exotic lines (Table 7.1). More recently, the Mtwapa program has released a hybrid, Pwani, intended for the sandy-soil areas of the humid lowlands near the coast.

A short time ago, Kenya established another maize breeding program at Kakamega in western Kenya, in recognition of the distinct biotic and abiotic stress factors and socioeconomic circumstances that characterize the relatively cooler areas of the moist midaltitude zone. The program's mandate is to develop maize materials for this environment, now referred to as the moist transitional zone (Chapter 4; Njoroge *et al.* 1992a).

Rates and Patterns of Adoption of Improved Maize Seed

Which maize varieties and hybrids developed by the Kenyan maize research program have been adopted by farmers? This section examines the patterns of adoption and measures the rates at which new maize varieties and hybrids spread among farmers in the different agroclimatic zones and farm-size groups. Sources of variation in adoption rates in different zones, for different types of maize seed, and among different farm-size classes are analyzed. It is important to note that the fact that seed of improved OPVs is usually recycled for several years makes it highly likely that, during the survey, the use of local materials was confused with the use of advanced generations of improved cultivars, leading to underestimation of the adoption of improved maize seed.

Adoption by agroclimatic zone and type of seed

Survey farmers in Kenya's six agroclimatic zones were asked to name the year in which they purchased improved seed for the first time. A logit trend curve[2] was fitted to these data to estimate rates of adoption – the slope parameter (b) – by zone. Adoption of improved maize was fastest among farmers in high-potential zones (the highland tropics and moist transitional zone), followed by the moist midaltitude and dry transitional zones. On the other hand, adoption was much slower in low-potential zones, such as the lowland tropics and dry midaltitude zone (Table 7.2 and Fig. 7.1).

The fact that adoption of improved maize seed was faster in the zones where more hybrids are grown and slower in areas where more OPVs are grown indicates the possible correlation between seed type and agroclimatic conditions. Table 7.2 and Fig. 7.2 show that farmers adopted hybrid seed more rapidly than seed of improved OPVs. Interactions between seed type and agroclimate are depicted in Figs 7.3–7.5, which show that late-maturing hybrids were rapidly accepted by farmers in the high-potential zones, whereas medium-maturing (500-series) hybrids, early-maturing composites, and maize developed for the coast were quickly adopted in the dry and moist midaltitude and lowland tropical zones, respectively.

The varied rates at which farmers in different zones adopt improved maize varieties and hybrids can be attributed to a number of factors: the rel-

Table 7.2. Rates of adoption of improved maize seed by zone and type of seed.

	Agroclimatic zone						Seed type		
	Lowland tropics	Dry midaltitude zone	Moist midaltitude zone	Dry transitional zone	Moist transitional zone	Highland tropics	Hybrid	Open-pollinated variety (composite)	Hybrid
Adoption rate (% yr^{-1}), log curve, 1960–1992[a]	0.12 (0.87)	0.16 (0.89)	0.19 (0.90)	0.18 (0.91)	0.23 (0.91)	0.26 (0.97)	0.23 (0.95)	0.14 (0.95)	0.24 (0.94)
First improved variety									
Name	Coast Composite	Katumani Composite A	Hybrid H511	Hybrid H511	Hybrid H611	Hybrid H611	Hybrid H611	Katumani Composite A	Hybrid H611
Year released	1974	1966	1968	1968	1964	1964	1964	1966	1964
Year 10% of farmers adopted	1983	1975	1972	1974	1968	1970	1970	1979	1970
Lag from release to 10% adoption (yrs)	9	9	4	6	4	6	6	13	6

[a] Estimates of the trend (slope) parameter (*b*) of the logistic adoption curve. Numbers in parentheses are R^2-values of the fitted trend equation. All parameter estimates were significant at 99%.

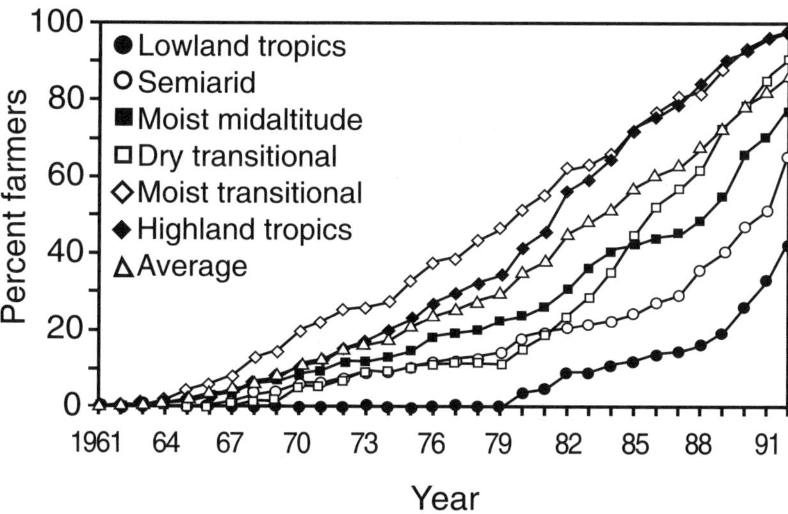

Fig. 7.1. Adoption of improved maize seed in Kenya, 1961–1992.

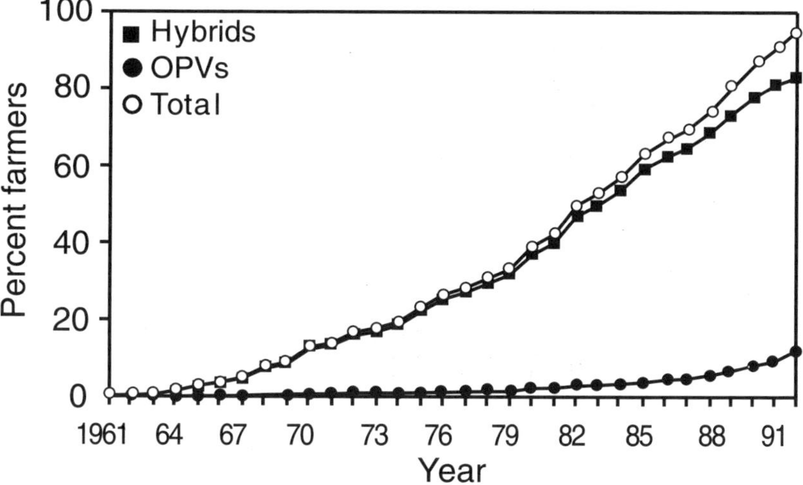

Fig. 7.2. Adoption of improved maize in Kenya by type of seed, 1961–1992.

atively later introduction of some kinds of improved seed, the information supplied to farmers about improved seed, the actual availability of the seed, the economic gains provided by improved seed, and other seed qualities (such as suitability for processing into local food preparations). First we will examine each of these factors in the Kenyan context; then we will assess adoption of improved seed by agroclimatic zone.

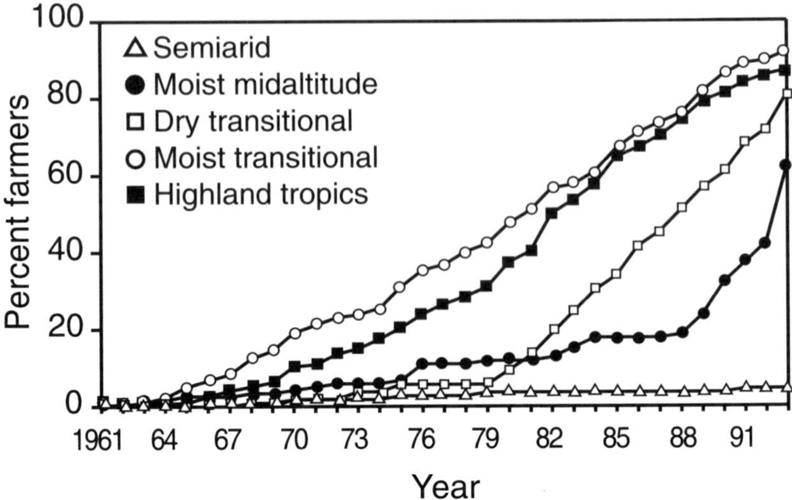

Fig. 7.3. Adoption of late-maturing maize (600-series) hybrids in Kenya, 1961–1993.

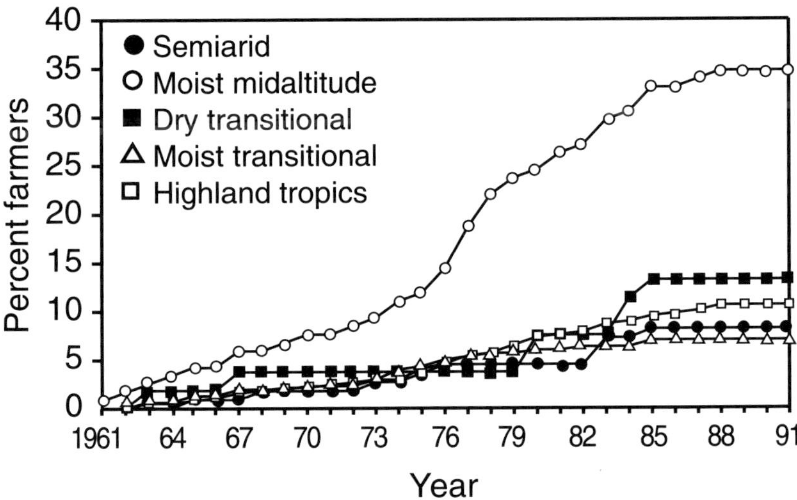

Fig. 7.4. Adoption of medium-maturing maize (500-series) hybrids in Kenya, 1961–1991.

Later introduction

Maize hybrids, especially the late-maturing (600-series) hybrids, were introduced into their target environments earlier than the composites were introduced. However, this does not explain the very large difference between hybrids and composites in the period between release and initiation of adoption (the year when 10% of the farmers adopt) (Table 7.2).

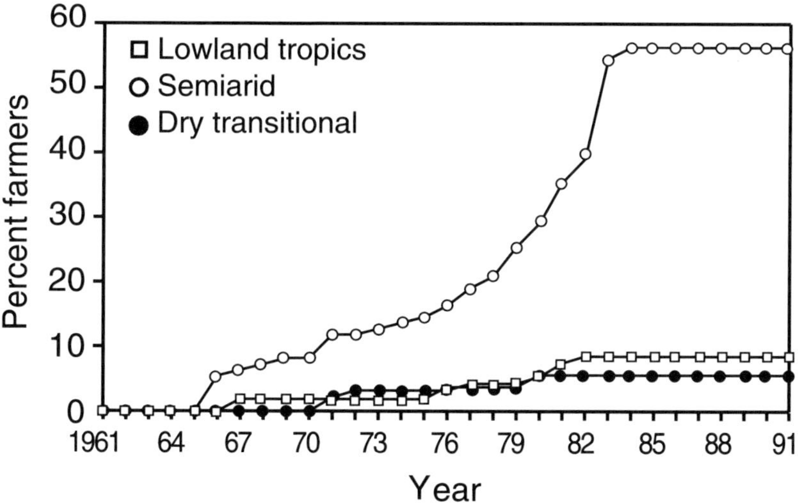

Fig. 7.5. Adoption of early-maturing maize germplasm in Kenya, 1961–1991.

Information and awareness

Farmers' awareness of the new technology is by and large a function of extension advice and observation or passive learning. Farmers in the high-potential areas where adoption was faster (the highland tropical and moist transitional zones) enjoyed better access to extension services than farmers in areas where adoption was slower, such as the dry midaltitude, dry transitional, and lowland tropical zones (Table 7.3). Moreover, the percentage of farmers indicating that they were not aware of the new seed was much higher in zones where adoption was low (Table 7.3).

Access to and availability of seed

Farmers' ability to obtain improved maize seed is mainly determined by the efficiency of the seed supply infrastructure and the availability of credit for purchasing seed. Access to improved seed was relatively poor in areas of low adoption compared with areas of high adoption (Table 7.3), where more seed outlets were located. Also, farmers in high-adoption areas had better access to credit than farmers in low-adoption areas (Table 7.3). Recent trends in the production and marketing of maize seed in Kenya are summarized in Appendix E. The emphasis on certain seed types, especially hybrids, is evident.

Profitability of new varieties

The most important factor in the diffusion of improved varieties is the economic incentive for farmers to use them. The following elements determine the profitability of improved maize seed:

- *Seed costs.* Since seed of all improved varieties, hybrids as well as OPVs, is sold at the same price in Kenya (Appendix E), price alone makes no

Table 7.3. Access to extension and the seed supply infrastructure in each agroclimatic zone.

Agroclimatic zone	Percentage of sites having extension agent based within survey site	Reason for not adopting improved seed (% farmers)				Percentage of sites having no seed outlet	Percentage of farmers used credit to buy seed	Road to nearest market		Adoption rate
		Farmer not aware	Seed not available	Expensive	Other			Percentage of roads are tarmac	Distance to market (km)	
Lowland tropics	20	43	27	10	20	80	1	0	9	0.12
Dry midaltitude	0	18	26	13	43	100	1	22	10	0.11
Moist midaltitude	25	4	31	22	43	75	1	12	18	0.19
Dry transitional	0	10	43	30	17	50	0	40	10	0.18
Moist transitional	44	6	17	10	67	41	7	23	7	0.23
Highland tropics	51	12	16	5	67	39	10	44	10	0.26
Average	33	12	22	12	54	56	6	28	10	

difference to adoption. However, seed becomes more expensive to acquire when transportation costs are high. Farmers in areas where adoption is low must travel longer distances on poorer roads to buy seed, unlike farmers in areas where adoption is high (Table 7.3). A negative relationship is revealed in Fig. 7.6 between distance from the farmer's source of maize seed and adoption of improved maize seed.[3]

- *Returns to use of new varieties.* Like maize seed, all maize grain is sold at the same price irrespective of its type. Farmers in areas where adoption of improved maize is low have difficulty marketing maize because of transportation infrastructure is relatively poor. In Kenya, the maize seed:grain price ratio fluctuated around a value of 3 between 1982 and 1993 (Appendix E), which is considered relatively favorable compared with price ratios in other maize-producing countries (CIMMYT 1994). It is important to note, however, that in areas where adoption of improved seed is low, maize is produced mainly for home consumption on very small holdings. Large commercial producers are concentrated in high-potential zones where access to markets is much better and the road network more developed. Although the Kenya Seed Company sought to accommodate smallholders by making seed available in smaller amounts (2 kg packages) during the 1980s, this trend was reversed in the early 1990s in favor of the medium-sized 10-kg seed packages, which account for more than 80% of all the seed packages sold (Appendix E).

Because all Kenyan maize farmers pay the same price for all types of maize seed and are paid the same price (set by the government) for whatever kind of maize they produce, variations in seeding rate and grain yield are considered the major determinants of comparative advantage in maize production.

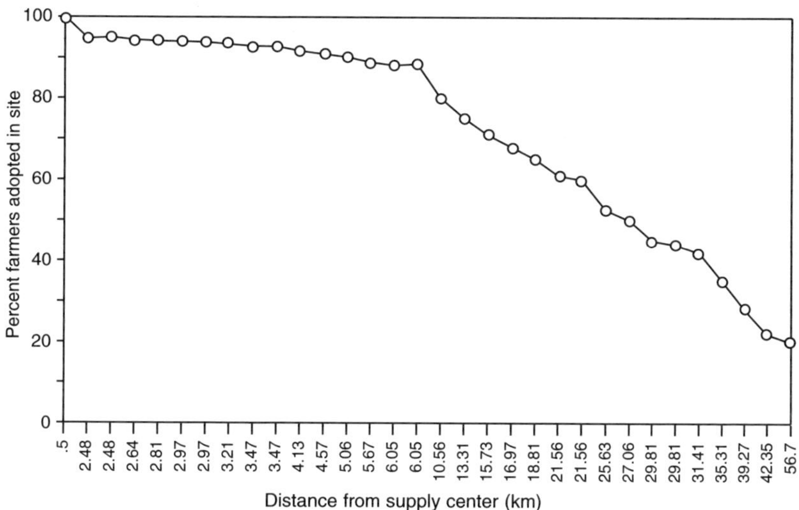

Fig. 7.6. Adoption of improved maize seed and distance from supply centers, Kenya, 1992.

Table 7.4. Average yield of maize cultivars (t ha^{-1}), by agroclimatic zone and level of soil fertility management.

	Unimproved maize cultivars		OPVs (composites)		Hybrids	
	No fertilizer	Basal fertilizer	No fertilizer	Basal fertilizer	No fertilizer	Basal fertilizer
Agroclimatic zone						
Lowland tropics	0.62	–	0.36	0.44	0.11	–
Dry midaltitude	0.28	–	0.64	0.42	0.65	1.28
Moist midaltitude	1.02	0.77	2.01	2.24	0.98	1.49
Dry transitional	0.43	0.59	–	–	0.73	1.02
Moist transitional	0.49	1.07	–	–	1.13	2.25
Highland tropics	1.70	1.89	–	–	1.63	2.29
Total	0.71	0.98	0.78	1.01	1.26	2.15
Number of farmers	420	56	54	2	253	619
Percentage of total	30	4	3.8	0.2	18	44
Percentage gain in yield						
From fertilizer application	–	38	–	30	–	70
From improved seed						
Without fertilizer	–	–	10	–	78	–
With fertilizer	–	–	–	3	–	120

Note: yield data represent farmers' average yield in the major maize-growing season of the survey year (1992/93).
– = Not applicable.

Unfortunately, the farmer survey did not provide good information on farmers' seeding rates, and for this study, yield gains realized by the survey farmers (in the fields they managed) were used as a crude measure of profitability. However, comparisons made on the basis of yield differentials should be interpreted with caution, given that the source of our data is the farmer survey and that it was not possible to control for yield-influencing factors other than variety. A more rigorous analysis of yield differentials is presented in the next chapter, using data from research trials.

The yield gain from improved varieties is therefore considered the major determinant of economic advantage. Yields reported by farmers varied by seed type and by agroclimatic zone. The yield advantage of hybrids over unimproved varieties was much larger than that of improved OPVs over unimproved materials, especially in high-potential zones (Table 7.4). While many farmers applied no fertilizer to hybrids, they still realized yields 78% higher than those obtained from unimproved local varieties. This differential was much higher than the 10% yield gain obtained by using improved OPVs rather than unimproved local materials. When fertilizer was applied to hybrids, the resulting yield gain was much higher (120%) than that of OPVs (3%) grown with fertilizer (Table 7.4). These results may explain the much higher adoption rates of hybrids, particularly in high-potential zones, compared with improved OPVs. The large differential in yield gains of hybrids and improved OPVs over unimproved cultivars raises questions about whether all types of improved maize seed should continue to be sold at the same price in Kenya.

Other seed qualities

Apart from their economic (yield) advantage, several other qualities may determine which varieties farmers choose to plant. Among the many attributes desired by farmers, the most important in food crops such as maize are taste, ease in processing, and storability. To improve the suitability of new varieties they develop, maize breeders must understand the importance of these traits to farmers and incorporate desirable traits into new germplasm.

During the survey, farmers were asked which maize variety was best for each of several traits, and varieties were ranked (Tables 7.5 and 7.6). Two schemes were then used to derive an overall ranking within each zone:

Scheme 1: The overall rank is derived as the simple sum of the ranks for all traits. This scheme assigns equal weights to all traits.
Scheme 2: In this scheme, it was assumed that agronomic qualities, such as tolerance to stress (i.e., low levels of moisture, low soil fertility, and late rains), are captured in the yield advantage. Therefore, an overall ranking was derived as a simple sum of the ranks in only yield and postharvest qualities, such as taste, processing, and storability. Equal weights are assigned to all traits in this scheme.

Lowland tropics

While farmers preferred Katumani Composite (KCB) for its yield advantage, they found local varieties superior to KCB in taste, processing, and storage qualities (Table 7.5). On the other hand, KCB was thought to perform better than local varieties under the stress conditions listed earlier. The other improved germplasm grown in this zone (Coast Composite and Pwani, grown by farmers near the coast) ranked third in all traits. Under Scheme 1, KCB and local cultivars were ranked at the same level, though this ranking was not consistent with the percentage of farmers who actually used the two kinds of varieties (coast locals ranked much higher than KCB) (Table 7.5). Under Scheme 2, which only considered yield and postharvest qualities, the ranking better matched farmers' actual choice of variety.

These results suggest that postharvest qualities are more important to farmers than yield, especially when the yield advantage over other materials is as low as it is for KCB in the lowland tropical zone. However, it is important to remember that poor access to information, credit, and seed are major constraints to the adoption of improved seed in this zone, especially at the coast, which lacks a seed distribution network, extension services, and marketing infrastructure. Even though the survey data are likely to underestimate the adoption of improved maize seed, it is still likely that improved postharvest qualities, larger yield gains, and more efficient seed supply, extension, and marketing services are crucial for wider adoption of improved maize varieties at the coast.

Table 7.5. Maize cultivars ranked according to assessments of various traits by farmers in the lowland tropics, dry midaltitude zone, and moist midaltitude zone.

	Yield		Taste		Storability		Processing		Overall ranking (Scheme 2)[a]	Water stress		Low soil fertility		Late rains		Overall ranking (Scheme 1)[a]	
Zone and cultivar	Rank	(%)	Rank	(%)	Rank	(%)	Rank	(%)		Rank	(%)	Rank	(%)	Rank	(%)	Rank	(%)[b]
Lowland tropics																	
Coast Composite	3	(9)	3	(6)	3	(5)	3	(13)	12	3	(6)	3	(0)	3	(0)	21	(8)
Katumani Composite	1	(52)	2	(33)	2	(14)	2	(40)	7	1	(50)	1	(50)	1	(50)	10	(9)
Coast Local	2	(30)	1	(56)	1	(76)	1	(47)	5	2	(44)	1	(50)	2	(43)	10	(78)
Hybrid Pwani	3	(9)	3	(6)	3	(5)	4	(0)	13	4	(0)	3	(0)	3	(0)	23	(5)
Dry midaltitude																	
Katumani Composite	2	(31)	2	(27)	2	(31)	2	(25)	8	1	(68)	1	(52)	1	(65)	11	(27)
Machakos Local	1	(35)	1	(40)	1	(36)	1	(37)	4	3	(11)	2	(29)	3	(11)	12	(51)
Hybrid H511	3	(8)	3	(5)	3	(6)	4	(12)	13	4	(8)	4	(7)	4	(6)	25	(4)
Hybrid H512	4	(7)	3	(5)	4	(4)	3	(14)	14	2	(12)	3	(9)	2	(13)	21	(4)
Makueni Composite	5	(2)	5	(0)	5	(3)	5	(3)	20	5	(1)	5	(2)	5	(4)	35	(1)
Moist midaltitude																	
Local white (Rachar)	1	(18)	1	(34)	1	(43)	3	(15)	6	1	(57)	1	(56)	1	(53)	9	(30)
Local yellow (Nyamula)	1	(18)	2	(17)	3	(13)	1	(24)	7	3	(9)	2	(12)	4	(8)	16	(17)
Hybrid H511	1	(18)	3	(12)	5	(6)	2	(19)	11	2	(10)	4	(7)	2	(12)	19	(11)
Hybrid H614	4	(15)	3	(12)	2	(15)	3	(15)	12	4	(7)	3	(11)	5	(5)	24	(12)
Hybrid H622	5	(13)	3	(12)	3	(13)	5	(12)	16	5	(5)	5	(5)	3	(9)	29	(11)
Hybrid H512	6	(8)	7	(3)	7	(1)	6	(6)	26	6	(4)	7	(2)	5	(5)	44	(3)
Hybrid H625	7	(7)	6	(6)	5	(6)	7	(5)	25	6	(4)	6	(4)	7	(3)	44	(9)

Note: Figs in parentheses denote the percentage of farmers indicating that the variety is the best for that respective trait.

[a] Overall ranking Scheme 1 is simply the sum of the ranks in all traits. Scheme 2 ranks are sums of the first four traits only. Lower values indicate higher rank (i.e., the best rank has a value of 1).

[b] The percentage of farmers who planted the variety in the 1992/93 season is given in parentheses.

Table 7.6. Maize cultivars ranked according to assessments of various traits by farmers in the dry transitional zone, moist transitional zone, and highland tropics.

	Yield		Taste		Storability		Processing		Overall ranking	Water stress		Low soil fertility		Late rains		Stalk lodging		Overall ranking (Scheme 1)[a0]	
Zone and cultivar	Rank	(%)	Rank	(%)	Rank	(%)	Rank	(%)	(Scheme 2)[a]	Rank	(%)	Rank	(%)	Rank	(%)	Rank	(%)	Rank	(%)[b]
Dry transitional																			
Hybrid H511	1	(36)	1	(29)	3	(17)	1	(25)	6	1	(39)	2	(25)	2	(25)	1	(45)	12	(22)
Hybrid H512	2	(25)	3	(19)	1	(22)	1	(25)	7	4	(11)	4	(17)	4	(0)	3	(18)	22	(12)
Machakos Local	3	(14)	2	(23)	5	(9)	4	(13)	14	3	(17)	2	(25)	1	(50)	4	(0)	24	(32)
Upland Local	4	(11)	5	(10)	3	(17)	3	(21)	15	5	(0)	5	(0)	4	(0)	4	(0)	35	(31)
Hybrid H64	5	(3)	4	(13)	1	(22)	5	(4)	13	5	(0)	5	(0)	4	(0)	4	(0)	33	(2)
Katumani Composite	6	(0)	6	(0)	6	(4)	5	(4)	23	2	(28)	1	(33)	2	(25)	2	(36)	34	(1)
Moist transitional																			
Hybrid H614	1	(39)	1	(41)	1	(44)	1	(44)	4	1	(30)	1	(40)	1	(24)	1	(67)	10	(40)
Hybrid H625	2	(31)	2	(22)	2	(24)	2	(17)	8	2	(17)	2	(13)	2	(17)	2	(17)	17	(27)
Hybrid H626	3	(8)	5	(4)	4	(4)	6	(4)	18	7	(4)	5	(3)	7	(3)	2	(17)	41	(12)
Hybrid H613	4	(5)	4	(6)	3	(8)	3	(8)	10	7	(4)	5	(3)	7	(3)	4	(0)	39	(1)
Upland Local	7	(2)	6	(2)	7	(2)	6	(2)	26	5	(5)	4	(7)	4	(9)	4	(0)	45	(4)
Hybrid H511	6	(4)	6	(2)	7	(2)	3	(8)	22	3	(11)	3	(8)	2	(17)	4	(0)	36	(4)
Hybrid H512	4	(5)	6	(2)	4	(4)	3	(8)	17	3	(11)	4	(7)	5	(8)	4	(0)	35	(3)
Highland tropics																			
Hybrid H614	1	(50)	1	(51)	1	(53)	1	(56)	4	1	(37)	1	(47)	1	(33)	2	(18)	9	(50)
Hybrid H625	2	(26)	2	(21)	2	(26)	2	(18)	8	2	(20)	2	(19)	2	(24)	3	(13)	17	(21)
Hybrid H626	4	(5)	4	(6)	5	(2)	5	(3)	18	6	(3)	5	(2)	6	(3)	4	(0)	39	(16)
Hybrid H613	3	(10)	3	(12)	3	(9)	3	(9)	12	3	(13)	3	(10)	4	(10)	4	(0)	26	(3)
Hybrid H622	5	(3)	5	(0)	4	(3)	5	(3)	19	6	(3)	5	(2)	6	(3)	4	(0)	40	(0)
Hybrid H511	6	(0)	5	(0)	6	(0)	4	(6)	21	4	(9)	4	(8)	3	(13)	1	(20)	33	(4)

Note: Numbers in parentheses denote the percentage of farmers indicating that the variety is the best for that respective trait.

[a] Overall ranking Scheme 1 is simply the sum of the ranks in all traits. Scheme 2 ranks are sums of the first four traits only. Lower values indicate higher rank (i.e., the best rank has a value of 1).

[b] The percentage of farmers who planted the variety in the 1992/93 season is given in parentheses.

Dry midaltitude zone

In this zone, as in the lowland tropics, KCB was regarded as the best variety under stress, whereas local cultivars were deemed superior for their postharvest qualities (Table 7.5). Other improved varieties, such as hybrids H511 and H512 and Makueni Composite, were the least valued by farmers. The overall ranking of Scheme 2 was more consistent with actual farmers' choices, compared with the Scheme 1 ranking. Once more, this result implies that the low yield advantage of KCB over local varieties in low-potential areas is less than adequate to compensate for the superiority of local cultivars in qualities other than yield. Note, however, that the same caveats that apply to recycling improved OPV seed and poor access to seed supply, extension, and credit in the lowland tropics hold for the dry midaltitude zone as well.

Moist midaltitude zone

Although the hybrid H511 ranked equal to local varieties in terms of yield, local maize dominated H511 with respect to nearly all other traits (Table 7.5). Again, the ranking in Scheme 2 was more consistent with farmers' choices than the ranking in Scheme 1. Local varieties were also preferred to improved cultivars for their better tolerance to the parasitic weed, *Striga*, the most important biotic stress factor in this zone (Hassan *et al.* 1994; Chapter 5). This preference was revealed here in farmers' assertion that local varieties tolerated low soil fertility better than other materials (low soil fertility and *Striga* infestation are correlated).

Dry transitional zone

Even though maize farmers in this zone indicated that the medium-maturing hybrids H511 and H512 yielded better than local varieties (Table 7.6), most farmers continued to grow local landraces. Again, the poor seed supply and the low yield advantage of these hybrids, especially without fertilizer (Table 7.4), appear to limit adoption in this zone. A high risk is associated with the use of inorganic fertilizers in this environment, where rainfall is erratic and other climatic risks high. Katumani Composite B retained its superior rank under stress conditions in this zone (Table 7.6).

Moist transitional zone and highland tropics

In these zones, where adoption of improved materials is high, the hybrid H614 was the variety most preferred for all traits (Table 7.6). Farmers in these zones clearly found late-maturing hybrids to be superior to local cultivars.

Adoption by farm size

Farmers' adoption behavior is influenced in many ways by the size of their land holdings, depending on the characteristics of the technology (Feder *et al.* 1985). For instance, lumpy technologies that involve large initial capital outlays (such as machinery, pumps, etc.) are adopted faster by large-scale farmers because they have relatively better access to capital and credit and the capacity to bear risk (Binswanger 1978; Gafsi and Roe 1979). However, even

Table 7.7. Adoption of improved maize varieties, by farm-size class.

	Large farms (> 1 ha maize)	Small farms (< 1 ha maize)
Adoption rates (log trend)[a]		
1960–1992	0.20 (0.90)	0.23 (0.97)
1960–1988	0.23	0.24
1960–1975	0.35	0.32
Year 10% of farmers adopted	1966	1972
Year 30% of farmers adopted	1969	1980
Percentage of farmers who never received extension advice	27	41
Percentage of farmers who used credit to buy seed	29	2
Farmers' reasons for not using improved maize seed		
Not aware of improved materials	6	14
Seed not available	9	25
Seed too expensive	6	13

[a] Numbers in parentheses are R^2 values of the fitted trend equation.

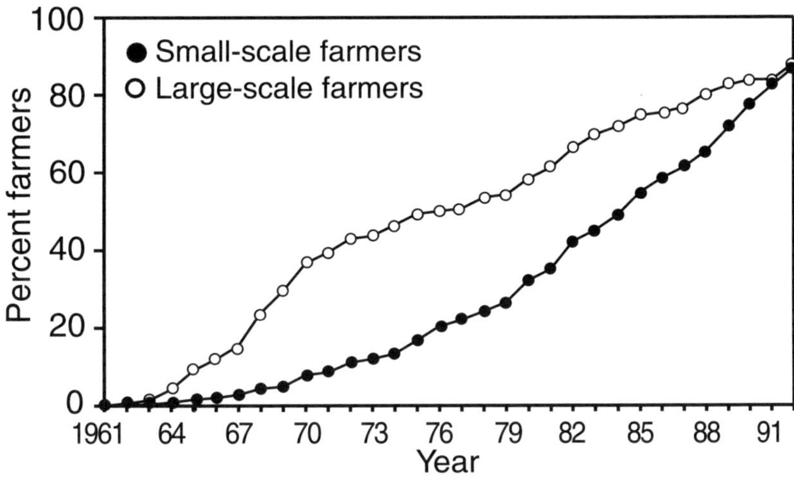

Fig. 7.7. Adoption of improved maize in Kenya by farm size, 1961–1992.

in the case of scale-neutral technologies such as modern varieties and fertilizers, small-scale farmers were found to lag behind other farmers in adoption, at least initially (Feder *et al.* 1985).

Results of the present investigation indicate that large-scale farmers were indeed early adopters of new maize varieties in Kenya (Table 7.7, Fig. 7.7). The average lag between the time of release of a variety and its initial adoption (i.e., by 10% of the farmers) was much shorter among large-scale farmers (two years) than smallholders (eight years). Even though small-scale farmers

Table 7.8. Farmers' use and sources of improved maize seed, by agroclimatic zone.

Agroclimatic zone	Percentage of farmers ever brought improved seed	Use of improved seed, 1992/93[a]		Percentage of farmers used own seed	
		Percentage of farmers used	Percentage of area planted	Hybrids	OPVs (composites)
Lowland tropics	43	16	16	20	39
Dry midaltitude	67	38	32	10	37
Moist midaltitude	79	41	51	0	25
Dry transitional	91	21	12	27	–[b]
Moist transitional	98	85	94	2	–[b]
Highland tropics	98	86	95	3	–[b]
Total	87	68	73	3	37

[a] Percentage of farmers who planted improved maize in the survey year (1992/93). Percentage of area refers to the percentage of area planted with improved maize seed.
[b] Very few, if any, farmers used composite (open-pollinated) varieties in these zones.

were slower to adopt a new variety, they eventually achieved adoption rates similar to those of other farmers (Fig. 7.7). Many studies of the relationship between farm size and adoption behavior report similar findings (Ruttan 1977). However, access to information, extension advice, and credit, which is far better for large-scale farmers in Kenya than for smallholders, may explain differential rates of adoption in the two groups (Table 7.7).

Disadoption and varietal replacement rates

The cumulative percentage of farmers who had ever bought improved seed was compared with the percentage of farmers who used improved seed in the survey year, 1992/93. The percentage of farmers using improved varieties in 1992/93 was consistently lower in all zones (Table 7.8), suggesting that some farmers who had once used seed of improved varieties no longer did so. The rate of disadoption appears to be lower in the high-potential zones, where only hybrids are used (the highland tropics and moist transitional zone). Table 7.8 also shows that farmers only infrequently plant advanced generations of hybrid maize: on average, only 3% of hybrid users did not buy new seed in 1992. On the other hand, a high proportion (37%) of farmers who plant improved OPVs (composites) used their own seed, saved from previous crops (i.e., advanced generations of the variety). The higher rates of disadoption in areas where farmers plant composites may be attributed to frequent use of advanced generations of improved OPVs. The seed of improved OPVs retains its genetic integrity longer than seed of hybrid maize, so OPV seed does not have to be replaced every year, and only a proportion of farmers growing composites buy seed in a given year. It is possible that some farmers have been recycling an improved variety for several years, and probably some farmers have never replaced the first seed they bought.

A measure of the rate of varietal replacement in farmers' fields was derived to examine the efficacy of the maize breeding program, the seed supply system, and extension services in maintaining a continuous flow of newer varietal releases for farmers to adopt. The present study measured the rate of varietal replacement among maize farmers in Kenya by applying the weighted average age (WAA) index proposed by Brennan and Byerlee (1991):

$$WAA_t = P_{it} R_{it}, \tag{7.1}$$

where P_{it} is the proportion of the area sown to variety *i* in year *t* and R_{it} is the number of years since the release of variety *i*. The WAA index was calculated for three categories of improved maize varieties grown in the 1992/93 season and for two farm-size groups (Table 7.9).

Late-maturing varieties (600-series hybrids) targeted for the high-potential zones had the smallest WAA, indicating rapid varietal turnover. One reason that farmers in these zones replace hybrids relatively more rapidly is because a large number of late-maturing varieties are released for these areas, compared with the medium- and early-maturing materials developed for other areas. The relatively better seed supply system, transportation network, and extension services in the high-potential zones (Table 7.3) also contributed to the rapid rate of varietal replacement. Large-scale farmers, because of their relatively better access to information, extension services, credit, and capacity to bear risk, achieved faster replacement rates compared with small-scale farmers (Table 7.9).

Table 7.9. Weighted average age and number of improved varieties and hybrids released in Kenya, 1960–1992.

Type of improved maize	Weighted average age in 1992 (years)	Number of varieties/hybrids released: Total	In past 20 years	In past 6 years	Number of varieties/ hybrids used in 1992/93
OPVs (composites)	23.2	4	2	1	3
Hybrids					
Medium-maturing (500 series)	23.3	2	0	0	2
Late-maturing (600 series)	9.0	14	7	4	6
Overall	10.3	23[a]	10[a]	6[a]	12
On large farms (> 1 ha)	9.2	–	–	–	–
On small farms (< 1 ha)	10.9	–	–	–	–

[a] The difference between the 'overall' totals and the sum of releases in the three germplasm categories is the result of excluding synthetic varieties and the hybrid Pwani, which do not belong to any of the categories listed here (see Appendix D).

Although genetic diversity of the maize materials planted in farmers' fields may widen over time if farmers replace older varieties with newer ones that incorporate new genes, spatial diversity within a farmers field is almost nonexistent, because few farmers mix varieties on the same farm. It is important to note that even if rapid varietal turnover in the 600-series hybrids implies widening genetic diversity in the varieties grown by farmers, the parentage of the other categories of improved seed (the 500-series hybrids and composites) is actually more diverse (Table 7.1).

The almost complete adoption of modern maize varieties in the high-potential zones of Kenya, coupled with rapid varietal replacement, indicate that these zones have entered the 'post-Green Revolution phase' of technical change (Byerlee 1992, 1994b). During this phase, yield gains resulting from the use of modern varieties are increased, or at least stabilized, through continued replacement of older modern varieties by newer generations of improved germplasm. On the other hand, maize production in the low- to medium-potential areas has not entered the post-Green Revolution phase; for these farmers, the Green Revolution itself is far from complete. Adoption of modern varieties remains very low, and varieties are replaced only very slowly. These variations in technical change at the farm level have important implications for the maize research priorities and strategies for the different zones. Technologies and strategies to promote input intensification and efficient input use are needed for achieving further productivity gains in the high-potential zones, but in zones of lower potential, institutional factors such as efficient seed supply and extension services, as well as faster varietal turnover, are crucial for attaining wider adoption and greater impact of improved varieties and production methods.

An Empirical Model of Farmers' Varietal Adoption Behavior

The previous analysis identified characteristics of maize hybrids and improved and local varieties that influence farmers' choice of maize germplasm. To test hypotheses about other factors that influence farmers to adopt modern seed technology, a formal model was developed that analyzes the effects on adoption of various agroclimatic and institutional factors as well as farmer characteristics. The following paragraphs describe the components of the model.

Agroclimatic conditions

Prevailing temperature and moisture regimes influence the growth and development of maize, which in turn determines the potential and suitability of improved maize in a given environment. Certain environments are more favorable for maize production than others, and hence larger gains can be achieved in these zones from improvement research. Stress factors vary by environment, and different biotic and abiotic stresses place different limits on a variety's ability to express the gains obtained through crop improvement research. The effects of some stress factors, such as erratic rainfall, heat, low soil fertility, insect pests, and weeds, are more difficult to alleviate through breeding than the effects of others, such as diseases. Some environments are also more favorable for certain types of seed than others. Hybrids appear to be more common in high-potential zones, whereas OPVs usually seem to be preferred in marginal environments. Zonal dummies were constructed to model the effect of agroclimatic diversity on the adoption of improved maize and the type of seed chosen.

Infrastructure and institutional factors

Better access to extension, credit, and seed supply promote adoption. Three variables were constructed to capture the effects of these factors.

Extension – Data on whether or not a farmer received any extension advice defined the extension dummy variable (1 = yes, 0 = no).
Credit – A dummy variable (1 = those who used credit to buy seed, 0 = those who did not) was used as a proxy for access to credit.
Seed supply – Two variables were used to test hypotheses about availability of the improved seed: a dummy indicating whether a seed dealer was operating in the survey site, and a continuous variable measuring the distance to the nearest permanent market.

Farmer characteristics

Socioeconomic attributes of the head of the farming family (being the decision maker) are often correlated with adoption patterns. Hypotheses about the relationship between the choice of the maize variety to be planted and the age, gender, and level of education of the farmer are tested with this model. Dummy variables were used to represent variations in education (none versus some education) and gender (male versus female). The age variable was measured in years.

Better education is assumed to enhance farmers' ability to acquire, interpret, and use information, including information about agricultural innovations, and hence leads to earlier and faster adoption (Evenson 1974; Gerhart 1975; Huffman 1977). Age may mean that a farmer has accumulated better information through longer experience and experimentation, and hence age is thought to increase the likelihood of adopting new technologies (Jamison and Lau 1982). On the other hand, attitude towards risk may be affected by age; older farmers may be more conservative or resistant to change than younger ones. The gender of the farmer can also influence the decision to adopt a new maize variety, especially if technology attributes related to gender-specific roles on the farm are changed. For instance, female farmers are expected to place more importance on postharvest qualities (i.e., processing for food) of new varieties than men do, because women usually perform postharvest operations (Hassan and Salasya 1994).

Farm size

The effect of the size of a farmer's land holding on the decision to adopt improved maize varieties was modeled using a dummy variable grouping farms into large (>1 ha of maize) and small (< 1 ha of maize) holdings. To decide on the cutoff point that would best separate large-scale from small-scale farmers, we experimented with several points. Only at 1 ha under maize did the two categories of farmers show significant variation in their seed adoption practices.

The dependent (decision) variables

Adoption is modeled in various ways, depending on the nature of the decision variable. Qualitative response models, such as the logit and probit procedures, are commonly used when the adoption process is dichotomous. In these models, a functional relation between the probability of adoption and a set of regressors is estimated econometrically using the logistic distribution for the logit procedure or the normal distribution for the probit procedure. The logit–probit method investigates the effects of explanatory variables on the choice to adopt or not adopt, but it does not measure the degree or intensity of adoption (Feder *et al.* 1985). Alternative econometric procedures are also used to analyze quantitative adoption decisions, when information on the extent and intensity of adoption is available (e.g., data on the percentage of area planted to modern varieties, amount of fertilizer applied, etc.). The tobit model is commonly used in the case of continuous but limited (censored) dependent variables.[4]

Two decision variables are used to study farmers' adoption of modern maize varieties in Kenya. Both are modeled as binary choices: whether or not to adopt improved maize seed and whether to plant hybrids or OPVs. This procedure was chosen because only a negligible number of farmers mix maize types; almost all farmers plant all their maize area to a single material, and there is no need to investigate area allocations among materials.

In this model, farmers choose between local varieties and two types of improved seed (improved OPVs and hybrids). The decision problem is thus separated into two stages, with each stage represented by a separate equation. One equation models farmers choice between local and improved maize varieties (Y_1). The second equation analyzes adopters decision about which type (Y_2) of improved seed to use: hybrid or improved OPV. (Nonadopters are excluded from Y_2). The questions of what influences farmers' choice of improved seed is important to answer, so that maize researchers can determine what type of maize will be appropriate to the biophysical and socioeconomic circumstances of maize farmers.

When the dependent variable is dichotomous, the assumptions of the standard linear regression model about the distribution of the random error component, such as multivariate normality and independence of regressor and error terms, are violated, rendering ordinary least squares inappropriate

(Amemiya 1973; Maddala 1983). The decision to adopt improved maize seed is modeled in equation (7.2) using the logit method:

$$y = 1 / (1+\exp(Z)), \quad (7.2)$$
$$Z = X'\beta,$$

where y is the probability that improved seed (Y_1) or a given seed type (Y_2) will be adopted, X is the set of regressors, and β is the vector of model parameters. Because of the large number of data points (more than 1400 observations),[5] normality is assumed, and thus results of the probit procedure were expected to be similar to the logit model. Because the relationship specified in equation (7.2) is nonlinear, the model is estimated using the iterative maximum-likelihood procedure of the statistical package SPSS (Norusis 1991). Estimates of the model parameters (β) are given in Table 7.10, together with the measure of odds of adoption, exp(Z). The odds factor is derived as:

$$F^i = \text{prob. (adoption)/prob. (nonadoption)} = \exp(Z),$$

where the odds of adoption change with X when β^i (coefficient on regressor i) is equal to zero; then Z is zero, and F^i is equal to one, indicating factor i has no effect (i.e., that it does not change the odds of adoption). On the other hand, when β^i is greater than zero, then F^i will be greater than one, indicating that the odds of adoption increase with X^i. The reverse is true when β^i is less than zero.

Model 1: Adoption of improved maize seed (Y_1)

Superior statistical performance of the model is evident from the statistics on goodness of fit (Table 7.10). The effect of agroclimate was statistically significant and consistent with earlier results, showing that the probability of adoption is higher in the more favorable maize environments. On the other hand, farm size and access to markets, measured in terms of the distance to the nearest permanent market, were not significant. Unlike proximity to markets, availability of improved seed (i.e., supply) proved to be an important determinant of adoption, with higher probability of adoption associated with the presence of a seed dealer at the village. Similarly, access to credit increased the likelihood of adoption.

Although extension contact increased adoption, level of education did not, which suggests that extension substitutes for education, especially in rural societies where illiteracy is high or only primary schooling is available. The fact that improved seed technology is relatively simple to use, compared with crop management methods, which involve more complicated processes that need to be understood and executed (Byerlee 1992), could be another reason why extension advice was found to be superior to formal education (and sufficient by itself) in promoting adoption of improved seed. However, better information, through education and/or extension, tended to increase the odds of adoption; that is, exp(Z) being greater than one.

The gender of the farmer was an important factor in the adoption of mod-

Table 7.10. Parameter estimates of the logistic regression model of farmers' varietal adoption behavior.

Dependent variable	Y_1 (improved vs. local maize)			Y_2 (hybrid vs. OPV)		
Model χ^2	401***			584***		
Goodness of fit	978***			1061**		
Number of valid cases	912			994		
Exogenous variables	β[b]		Exp(Z)	β[b]		Exp(Z)
Agroclimatic zones[a]						
Semiarid	−0.81	(0.19)***	0.45	−1.59	(.27)***	0.21
Moist midaltitude	−0.31	(0.20)*	0.74	0.27	(.26)	1.31
Dry transitional	−1.31	(0.29)***	0.27	−0.22	(.28)	0.80
Moist transitional	1.76	(0.19)***	5.81	2.46	(.26)***	11.68
Highland tropics	2.27	(0.21)***	9.66	3.02	(.28)***	20.56
Farm size (large)	0.32	(0.49)	1.37	0.64	(0.39)*	1.89
Distance to nearest market (km)	−0.01	(0.007)	0.99	0.007	(0.008)	1.01
Seed dealer in the village	0.12	(0.09)*	1.13	0.24	(0.16)**	1.24
Availability of credit	0.9	(0.55)*	2.45	0.66	(0.53)	1.93
Extension advice (received)	0.4	(0.10)***	1.49	0.56	(0.10)***	1.75
Education (some)	0.024	(0.1)	1.03	0.13	(0.11)	1.14
Gender of farmer (male)	0.20	(0.1)**	1.23	0.23	(0.11)**	1.25
Age of farmer (years)	−0.03	(0.01)***	0.98	−0.03	(0.007)***	0.98
Cropping system (pure stand maize)	0.03	(0.11)	1.03	0.13	(0.12)	1.13
Constant	1.45	(1.01)	–	0.29	(0.99)	–

[a] To avoid singularity of the $x'x$ matrix, one dummy was dropped using the rule: $d_i = 1$, where d_i is the deviation of the mean of category i from the overall mean, i=1,2,..., $N-1$, and N is the total number of dummies. Accordingly, the coefficient on the missing dummy (i.e., lowland tropics) is equal to $\Sigma_i d_i$ for i $N-1$.

[b] Numbers in parentheses denote standard errors; *, **, *** indicate statistical significance at 10%, 5%, and 1%, respectively.

ern maize varieties in Kenya. Male farmers were more likely to adopt new varieties than female farmers. The fact that local varieties are preferred by farmers for their superior taste and processing qualities, especially in low-potential zones (see Table 7.8), may explain the negative relation between the probability of adoption and female farmers, since postharvest operations are mainly the women's job in rural Kenya (Hassan and Salasya 1994). On the other hand, farmer's age was statistically insignificant and negatively correlated with the probability to adopt. This result implies that conservatism increases with age as hypothesized.

Model 2: Choice of improved seed type (Y_2)

Goodness of fit measures indicate the high accuracy of the logistic model in forecasting observed responses in terms of calculated probabilities (Table 7.10). As expected, the likelihood of choosing a hybrid was much higher in high-potential areas, where their yield advantage is greatest (Table 7.4). The reverse was true for OPVs, which were preferred in marginal environments, especially the drier areas (the dry midaltitude and dry transitional zones) (Table 7.10). The odds of choosing a hybrid were higher among large-scale farmers and farmers who had better access to credit. The fact that hybrid seed must be purchased every year makes it more expensive to use than seed of improved OPVs, especially for small-scale farmers who cannot obtain credit. These results suggest that, in addition to the physical environment, financial constraints are an important influence on the diffusion of hybrid seed.

While access to markets was (again) unimportant, the presence of a seed dealer in the village increased the chances that hybrids would be adopted. This finding could be the result entirely of supply factors, if seed outlets stock only hybrids or have disproportionally higher stocks of hybrid seed compared with seed of OPVs. Aggregate data on total seed production and sales show that the share of improved OPVs was only 2%, on average, over 1987–1991 (Appendix E). The effect of extension advice was highly significant, indicating that extension contact increases the probability of adopting hybrid maize compared with improved OPVs (Table 7.10). This result may be partially explained by the fact that better extension services tend to be concentrated in areas where the physical climate is more favorable for hybrids (Table 7.3), but it also may suggest that extension promotes the use of hybrid maize.

Age and gender also affected the type of seed farmers selected. Hybrids were preferred more than OPVs by males and younger farmers (Table 7.10). Just as small-scale farmers may prefer OPVs to hybrids because they cannot obtain credit to purchase improved seed, the preference of women farmers for OPVs over hybrids may be related to financial considerations (hybrids are more expensive to use), as women are generally observed to have relatively poorer access to financial resources (Hassan and Salasya 1994). On the other hand, several factors may explain the age effect. Increased conservatism with age may be one reason for the negative correlation between age and probability of adoption of hybrid seed (the more expensive seed option). Younger farmers may also have better access to formal education and extension advice, which increased the odds that a farmer would adopt hybrids (Table 7.10). In addition, it is possible that the wife of an older male farmer will participate more actively in decision making, especially in managing the maize crop, and that like other women she will be biased toward OPVs. Although these correlations could not be established here, the high statistical significance of the age effect indicates that a more comprehensive investigation into the role of age in the adoption of improved maize is warranted.

Conclusion: Implications of the Study

Thirty years after hybrid maize began to spread rapidly into the high-potential zones of Kenya, the adoption of improved maize varieties and hybrids remains incomplete in many areas. The analysis presented here has examined which farmers plant improved maize varieties and hybrids, which ones do not, and the reasons for these choices.

The analysis has shown that adoption of modern maize varieties was much higher and faster in Kenya's high-potential zones compared with the dry midaltitude zone and lowland tropics. Hybrid maize was more rapidly accepted by farmers than improved OPVs. Several factors contributed to this adoption pattern. First, hybrids had a much higher yield advantage over local cultivars than improved OPVs, especially when fertilizer was used. The substantial yield advantage may explain the faster adoption of hybrids compared with improved OPVs, especially in low-potential areas where highly erratic rainfall increases the risk of using commercial fertilizer. On the other hand, the large differential in yield between hybrids and improved OPVs casts doubt on the advisability of selling both types of seed at the same price, which may discourage farmers from purchasing seed of improved OPVs.

Second, hybrids are most commonly grown in high-potential areas, where the seed distribution infrastructure and roads are much more developed than they are in low-potential areas. Better access to seed contributes to a higher level of adoption of hybrids and also to the higher rate of varietal turnover. The weighted average age of hybrids grown by farmers was nine years, compared with 23 years for OPVs. Not only do farmers in high-potential areas have better access to seed, they have a wider range of materials to choose from. Over the same period, only four improved OPVs were released in the low-potential zones compared with 16 hybrids released in the medium- to high-potential zones. The number of hybrids currently planted by farmers was larger than the number of OPVs. Although the wider range of hybrid maize materials planted in farmers' fields may represent a wider range of genetic diversity, it is important to note that the genetic background of these late-maturing hybrids is actually narrower than it is in the medium- and early-maturing varieties.

A third factor working against adoption of improved OPVs is their reported inferiority to local cultivars in desirable qualities other than yield, such as taste, processing, and storability. Fourth, farmers in high-potential zones had better access to credit and extension services than farmers in low-potential areas. Fifth, farm size was an important determinant of early adoption, as farmers with larger holdings adopted improved varieties only two years, on average, after their release. Even though smallholders eventually achieved adoption rates similar to those of large-scale farmers, the adoption process was much longer: on average, smallholders adopted new varieties eight years after release. Large-scale farmers enjoyed much better access to credit and information about modern varieties. Results of the logit regression indicated that the probability of adoption of improved maize materials, especially hybrids, increased in favorable environments. Better information, more education, access to credit and markets, larger holding size, and younger age were positively correlated with the probability that a farmer would adopt modern maize varieties, particularly hybrids. The likelihood of adoption was also higher among male farmers than female farmers.

Clearly, larger gains in yield and better postharvest qualities are needed if improved OPVs are to replace local cultivars in farmers' fields in low-potential zones. It is also clear that investment in improving seed distribution, extension, and the physical infrastructure in low-potential areas is crucial for wider adoption of improved maize seed technology. For instance, the quantity of OPV seed produced and marketed by the Kenya Seed Company is quite small compared with the quantity of hybrid seed. The research system must speed the development and release of a greater number of suitable OPVs to enhance adoption and productivity gains in low-potential zones and to achieve faster turnover. Also, seed pricing policies need to be revised to reflect not only the cost but, more importantly, the differential yield gains of different kinds of improved maize seed. Seed of OPVs appears to be overpriced compared with seed of hybrids.

Whereas lower adoption of improved maize, mainly because of supply and institutional factors, is the constraint to increased productivity of maize in low-potential areas, in the high-potential areas, where most farmers plant hybrids, significant gains in maize productivity can be achieved from increased use of fertilizers. A significant proportion of these farmers do not apply fertilizer, and consequently the hybrids do not express their true yield potential. Although crop management and improvement research can contribute to developing hybrid maize that performs well under conditions of low soil fertility, pricing and marketing policies can also contribute to improving the productivity of maize by promoting the use of commercial fertilizers. Better education and credit facilities may also increase smallholders' use of commercial fertilizers. Policies and institutional arrangements that would improve the access of younger farmers to land could also promote wider adoption of improved maize, especially hybrids, resulting in larger productivity gains.

References

Amemiya, T. (1973) Regression analysis when the dependent variable is truncated normal. *Econometrica* 41, 997–1016.

Binswanger, H. (1978) *The Economics of Tractors in South Asia: An Analytical Review.* Agricultural Development Council and the International Crops Research Institute for the Semi-arid Tropics (ICRISAT), New York.

Blackie, M. (1989) *Review of Maize in Eastern and Southern Africa.* The Rockefeller Foundation, Lilongwe, Malawi.

Brennan, J., and Byerlee, D. (1991) The rate of crop varietal replacement on farms: Measures and empirical results for wheat. *Plant Varieties and Seeds* 4, 99–106.

Byerlee, D. (1992) Technical change, productivity, and sustainability in irrigated cropping systems of South Asia: Emerging issues in the post-Green Revolution era. *Journal of International Development* 4, 477–496.

Byerlee, D. (1994a) *Maize Research in Sub-Saharan Africa: An Overview of Past Impacts and Future Prospects.* CIMMYT Economics Working Paper 94/03. CIMMYT, Mexico City.

Byerlee, D. (1994b) *Modern Varieties, Productivity, and Sustainability: Recent Experience and Emerging Challenges.* International Maize and Wheat Improvement Center (CIMMYT), Mexico City.

CIMMYT (International Maize and Wheat Improvement Center) (1994) *1993/94 CIMMYT World Maize Facts and Trends. Maize Seed Industries, Revisited: Emerging Roles of the Public and Private Sectors.* International Maize and Wheat Improvement Center (CIMMYT), Mexico City.

Dowker, B. (1964) New cereal varieties 1963: Katumani synthetic No. 2. *East African Agriculture* 30, 31–32.

Evenson, R. (1974) Research, extension, and schooling in agricultural development. In: Foster, P. and Sheffield, J. (eds) *World Year Book of Education.* Evans Brothers, London, UK.

Feder, G., Just, R., and Zilberman, D. (1985) Adoption of agricultural innovations in developing countries: A survey. *Economic Development and Cultural Change* 33(14), 255–298.

Gafsi, S., and Roe, T. (1979) Adoption of unlike high-yielding wheat varieties in Tunisia. *Economic Development and Cultural Change* 28, 119–134.

Gerhart, J. (1975) *The Diffusion of Hybrid Maize in Western Kenya.* Research report for CIMMYT. CIMMYT, Mexico City. Mimeo.

Griliches, Z. (1957) Hybrid corn: An exploration in the economics of technological change. *Econometrica* 25, 501–522.

Harrison, M. (1970) Maize improvement in East Africa. In: Leakey, C.L.A. (ed.) *Crop Improvement in East Africa.* Commonwealth Agricultural Bureaux, Farnham Royal, UK.

Hassan, R., and Salasya, P. (1994) The gender factor in maize production and technology transfer in Kenya. *Proceedings of the Fourth KARI Scientific Conference.* Kenya Agricultural Research Institute, Nairobi, Kenya.

Hassan, R., Ransom J., and Ojiem, J. (1994) The spatial distribution and farmers' strategies to control *Striga* in maize: Survey results from Kenya. In: *Proceedings of the Fourth Eastern and Southern Africa Regional Maize Conference.* International Maize and Wheat Improvement Center (CIMMYT), Harare, Zimbabwe.

Huffman, W. (1977) Allocative efficiency: The role of human capital. *Quarterly Journal of Economics* 91, 59–80.

Jamison, D.T., and Lau, L. (1982) *Farmer Education and Farm Efficiency.* Johns Hopkins University Press, Baltimore, Maryland.

Maddala, G.S. (1983) *Limited-dependent and Qualitative Variables in Econometrics.* Cambridge University Press, New York.

Mohamed, L., Scott, F., and Steeghs, H. (1985) Seed availability, distribution, and use in Machakos District. Ministry of Agriculture, National Dry Land Farming Research Center, Katumani, Kenya. Unpublished report.

Njoroge, K., Kanampiu, N., Otsyula, R., Muthamia, Z., Gathuri, C., and Chivatsi, W. (1992a) The high-altitude maize breeding program. In: *Proceedings of a Workshop on Review of the National Maize Research Program.* Kenya Agricultural Research Institute and the International Service for National Agricultural Research (ISNAR), Nairobi, Kenya, pp. 20–31.

Njoroge, K., Kanampiu, N., Otsyula, R., Muthamia, Z., Gathuri, C., and Chivatsi, W. (1992b) The Katumani mid-altitude maize breeding program. *Proceedings of a Workshop on Review of the National Maize Research Program.* Kenya Agricultural

Research Institute and the International Service for National Agricultural Research (ISNAR), Nairobi, Kenya, pp. 40–44.

Njoroge, K., Kanampiu, N., Otsyula, R., Muthamia, Z., Gathuri, C. and Chivatsi, W. (1992c) The Embu maize improvement program. In *Proceedings of a Workshop on Review of the National Maize Research Program.* Kenya Agricultural Research Institute and the International Service for National Agricultural Research (ISNAR), Nairobi, Kenya, pp. 37–39.

Njoroge, K., Kanampiu, N., Otsyula, R., Muthamia, Z., Gathuri, C. and Chivatsi, W. (1992d) The coast maize improvement program. In *Proceedings of a Workshop on Review of the National Maize Research Program.* Kenya Agricultural Research Institute and the International Service for National Agricultural Research (ISNAR), Nairobi, Kenya, pp. 35–36.

Norusis, M. (1991) *SPSS/PC+ Advanced Statistics 4.0 for IBM PC/XT/AT and PS/2.* SPSS, Chicago, Illinois.

Ogada, F. (1975) The effect of selection prolificacy in two composite populations of maize. PhD thesis, Makerere University, Kampala, Uganda.

Ongaro, W.A. (1991) Modern maize technology, yield variations, and efficiency differentials: A case of small farms in Western Kenya. *Eastern Africa Economic Review* 6, 11–30.

Ruttan, V. (1977) The green revolution: Seven generations. *International Development Review* 19, 16–23.

Sarch, M., and Gilbert. E., (1995) *Maize Research Impact in Kenya.* U.S. Agency for International Development, Bureau for Africa, Productive Sector Growth and Environmental Division, Maize Research Impact in Africa Project, Washington, DC.

USAID (US Agency for International Development) (1980) *Kitale Maize: The Limits of Success.* AID Project Impact Evaluation Report No. 2. US Agency for International Development, Washington, DC.

Endnotes

1 Both composite and synthetic varieties are made by mixing genotypes of known combining values for direct release to farmers. Synthetics, however, are presumed to have a narrower genetic base than composites. These two types of OPVs have also been used by breeders as sources of further genetic improvement in their breeding materials.

2 The log trend follows an S-shaped curve that represents three stages in adoption of new innovations: a flat segment representing an initial phase of hesitation and slow adoption; an intermediate phase of fast adoption (when more farmers learn about the advantages of the new innovation); and a final phase of slow adoption, when a small number of additional farmers adopt the innovation. The log trend curve traces cumulative percentage adoption using the formula $y_t = K/[1 + \exp(-a-bt)]$, where y_t is the percent adopting at time t and K is the maximum percent adoption or the ceiling rate (Griliches 1957). Based on maize researchers' and extension officers' estimates, a ceiling of 100% maximum adoption was assumed for zones where adoption exceeded 80%; a ceiling rate of 80% was assumed for zones with less than 80% but more than 50% adoption; and a ceiling rate of 60% was assumed for zones where adoption was lower than 50%.

3 A linear trend line was fitted to the original data to generate Fig. 7.6, which shows a slope of large magnitude (−1.83) and high statistical significance (1%).

4 See Feder *et al.* (1985) for a comprehensive survey of adoption studies.
5 Data on current varietal choices (e.g., adoption in the survey year) rather than on historical adoption patterns were used to construct the dichotomous choice variables (Y_1 and Y_2) for the logit analysis.

8 Determinants of Fertilizer Use and the Gap between Farmers' Maize Yields and Potential Yields in Kenya

RASHID M. HASSAN, FESTUS MURITHI, AND GEOFRY KAMAU

The rapid diffusion of maize hybrids in Kenya is often considered equivalent to the success of hybrid maize in the US, but fertilizer use has remained low, particularly among smallholders and in marginal environments. Data from a survey of 1400 farmers, along with five years of data from trials assessing maize response to fertilizer at more than 70 locations, were used to examine the current state and determinants of fertilizer use on maize and assess the gap between farmers' maize yields and potential yield levels. Kenyan farmers apply lower rates of inorganic fertilizers on their maize crop than is considered economically optimal. A serious imbalance in the ratio of applied N to P has led to mining of soil N. The gap between farmers' yields under current practices and the yields that could be attained if optimal levels of N and P were applied is considerable. About one million more tons of maize could be added to current domestic production (an increase of one-third) if farmers improved their soil fertility management practices. Two factors are crucial for increasing farmers' demand for fertilizer: high yield gains from fertilizer use and lower nutrient–grain price ratios. Among the supply side forces that influence the price of fertilizer are transportation and transaction costs. The other major supply factor is the dominance of low N fertilizers among the commercial products traded in Kenya. Promotion of high N products such as urea will reduce the relative cost of the N component, lower the nutrient–grain price ratio, and improve the returns to fertilizer use for maize farmers. Public investment in rural roads, removal of restrictions on fertilizer imports, and the provision of credit and tax incentives to private traders to invest in private storage and distribution facilities would have a significant impact on the diffusion of fertilizer. The marketing of fertilizer in smaller packages would also help increase fertilizer use, especially among small-scale farmers in remote areas. Yet another means of increasing the profitability of fertilizer use would be to achieve higher yield gains (biological response) from fertilizer application, both through breeding research (the development of better adapted hybrids for marginal zones) and crop management research and extension efforts (to disseminate fertilizer recommendations conditioned by soil type and nutrient analysis and to develop alternative low-cost sources of N).

Kenyan farmers knew about and used mineral fertilizers long before they used hybrid maize. With the release of hybrid maize in the late 1960s, fertilizer use on maize grew rapidly, especially on the large commercial farms in the high-potential areas where hybrid maize was first adopted (Gerhart 1975; G.M. Ruigu and M. Schluter, unpublished paper). By the mid-1970s, hybrid maize had become popular among smallholders as well, and its use spread to the medium-potential zones. However, while the rapid diffusion of maize hybrids in Kenya is often considered equivalent to the success of hybrid maize in the US, fertilizer use has remained low, particularly among smallholders and in marginal environments. For this reason, the increased use of inorganic fertilizer is considered an important source of potential productivity growth in maize in Kenya. This is particularly true in the high- and medium-potential zones, where more than 70% of Kenya's maize area is found and where the potential for productivity growth is largest (Chapter 6). Moreover, mining of soil nutrients is already very high in Kenya as a result of intensive farming and low application of fertilizer, and the balance between supply and extraction of nutrients is negative (Ransom *et al.* 1993; Smaling *et al.* 1993a; Smaling *et al.* 1993b; Stoorvogel *et al.* 1993).

In this chapter we analyze the current state and determinants of fertilizer use on maize and assess the gap between farmers' maize yields and potential yield levels in Kenya. First, we review the supply and demand forces affecting the diffusion of fertilizer for maize production. We move on to explore the variation in fertilizer adoption and potential productivity gains by agroclimate, soil type, and farmer groups, with the objective of generating useful information for proper targeting and conditioning of fertilizer recommendations. A final objective of the analysis is to identify the policy and institutional constraints, and consequently the reforms needed, to promote higher levels of fertilizer use among maize farmers.

Data Sources

Two primary sources were used in the present study. The MDBP survey (described in Chapter 3) gathered information from 1400 farmers on the levels, types, and amounts of fertilizer they applied to maize; when they first adopted inorganic fertilizer; where they obtained the fertilizer; their access to agricultural services supporting fertilizer use; and practices related to fertilizer use. Experimental records from the Fertilizer Use and Recommendations Project (FURP) were compiled on maize response to fertilizer by agroclimate, type of maize seed, cropping pattern, and soil type from more than 70 FURP sites over five years (KARI 1990). The MDBP agroclimatic zones and other spatial data were used as well.

Historical Background to Fertilizer Use on Maize in Kenya

Commercial fertilizers have been used in Kenya since the 1920s, at first mainly by large-scale white settler farmers who produced cash crops such as tea, coffee, and sugarcane (G.M. Ruigu and M. Schluter, unpublished paper; Olouch-Kosura and Chege 1994). After Independence in 1963, Kenyan farmers were allowed to enter coffee and tea farming and started using fertilizer. A fertilizer subsidy scheme introduced by the government, plus the release of high-yielding hybrid maize, encouraged fertilizer use among maize farmers during the 1960s (Allgood *et al.* 1990; Kimuyu *et al.* 1991; Murithi and Shiluli 1993).

The evolution of fertilizer use by maize farmers over the past three decades is shown in Table 8.1. Three periods are defined, representing different phases in the process of technological transformation in Kenya's maize subsector. Phase one (1964–1973) corresponds to the period when the first

Table 8.1. Fertilizer use by maize farmers in Kenya, 1964–1992.

	1964–73	1974–83	1984–92	Total
High-potential zones[a]				
Large-scale[b] (adoption rate: % yr^{-1})	6.7	2.2	0.9	3.3
Average kg nutrients used per ha	na	60[d]	98 (63%)	98
Small-scale[b] (adoption rate: % yr^{-1})	1.1	2.5	3.7	2.4
Average kg nutrients used per ha	na	10[d]	34 (260%)	36
Medium-potential zones[a]				
Small-scale[c] (adoption rate: % yr^{-1})	0.7	1.0	2.1	1.3
Average kg nutrients used per ha	na	13[d]	30 (130%)	30
Low-potential zones[a]				
Large-scale[b] (adoption rate: % yr^{-1})	0	0	0	0
Average kg nutrients used per ha	na	na	0	0
Small-scale[b] (adoption rate: % yr^{-1})	0.2	0.2	0.5	0.33
Average kg nutrients used per ha	na	na	11	11
Percentage farmers used fertilizer before improved seed	5	11	10	9
Average kg nutrients used per ha	na	24[d]	44 (83%)	44
Percentage of total nutrients applied on maize[e]	na	20	26	26

Note: reported amounts of fertilizer applied refer to the average for all farmers; na = data not available. Numbers in parentheses denote the percentage increase in nutrient use.

[a] High-potential zones consist of the highland tropics and moist transitional zones; medium-potential zones are the dry transitional and moist midaltitude zones; and low-potential zones comprise the semiarid zone and lowland tropics.

[b] Large-scale farmers have more than 8 ha of land.

[c] Only three farmers were found to have more than 4 ha, so the group of large-scale farmers was dropped from this category.

[d] Average rates applied in 1983 (Ruigu and Schluter unpublished paper).

[e] The 1983 share is extracted from CIMMYT (1992). Current (1992) share was derived by the authors based on total area under maize and total nutrients consumed in 1992 (Government of Kenya unpublished paper).

generation of hybrid maize materials was released and rapidly adopted by farmers in the high-potential zones.[1] The spread of improved maize seed was followed closely by the use of fertilizer on maize. This pattern is reflected in the much higher adoption rates for fertilizer during the initial phase among large-scale farmers in the high-potential zones where hybrid maize was first adopted.

Fertilizer is generally adopted by maize farmers at the same time as improved seed or afterwards (Table 8.1 and Fig. 8.1). As improved maize seed spread to small-scale farmers in the high- and medium-potential zones during the second phase (1974–1983), those farmers increased their use of fertilizer. During the third phase (1984–1992), small-scale farmers continued to adopt fertilizer at even faster rates than large-scale producers in the high-potential zones. The variation in the rates of diffusion of fertilizer by zone and farm size are depicted in Figs 8.2 and 8.3.

Fertilizer use may have spread rapidly in Kenya's more favorable maize production zones, but it spread very slowly in the low-potential zones. The high risk associated with fertilizer use in the marginal environments (Nadar and Faught 1984; Keating *et al.* 1993), especially on large areas, may be the reason for large-scale farmers' reluctance to apply fertilizer in these zones. It is also important to note that large commercial maize production in the marginal zones is quite insignificant.

Levels of nutrients applied to maize by smallholders in the high-potential zones increased at twice the rate of use in medium-potential areas during the past decade (Table 8.1). On average, the amount of nutrients applied to maize rose from 24 kg ha^{-1} in 1983 to 44 kg ha^{-1} in 1992 (an 83% increase). This rate compares favorably with those in most developing countries, especially in dryland agriculture, but it is much lower than fertilizer application rates in industrialized countries, where maize receives about 300 kg ha^{-1} (Larson 1993; Heisey and Mwangi 1997). Nevertheless, levels of nutrients used are still much lower than economically optimum rates (listed in Table 8.5).

While there is evidence that total use of fertilizer on all crops declined in the late 1980s in Kenya (Murithi and Shiluli 1993; Government of Kenya, unpublished report), the share of maize in total fertilizer use increased from 20% in 1983 to 26% in 1992 (Table 8.1). This result is consistent with findings of other studies, which have suggested that there was a general shift in fertilizer use to food crops in sub-Saharan Africa during the 1980s (Heisey and Mwangi 1997). It is also possible that some of the fertilizer provided at subsidized prices to tea and coffee farmers in Kenya has been finding its way to their maize fields (Kimuyu *et al.* 1991; Government of Kenya, unpublished report).

Demand and Supply Forces Affecting Fertilizer Use on Maize

Many factors influence the demand for and availability of fertilizer at the farm level, but it is not a simple task to distinguish between supply and demand forces. Here we have classified the role of marketing policies, distribution infrastructure, and packaging as supply factors. We consider other factors affecting

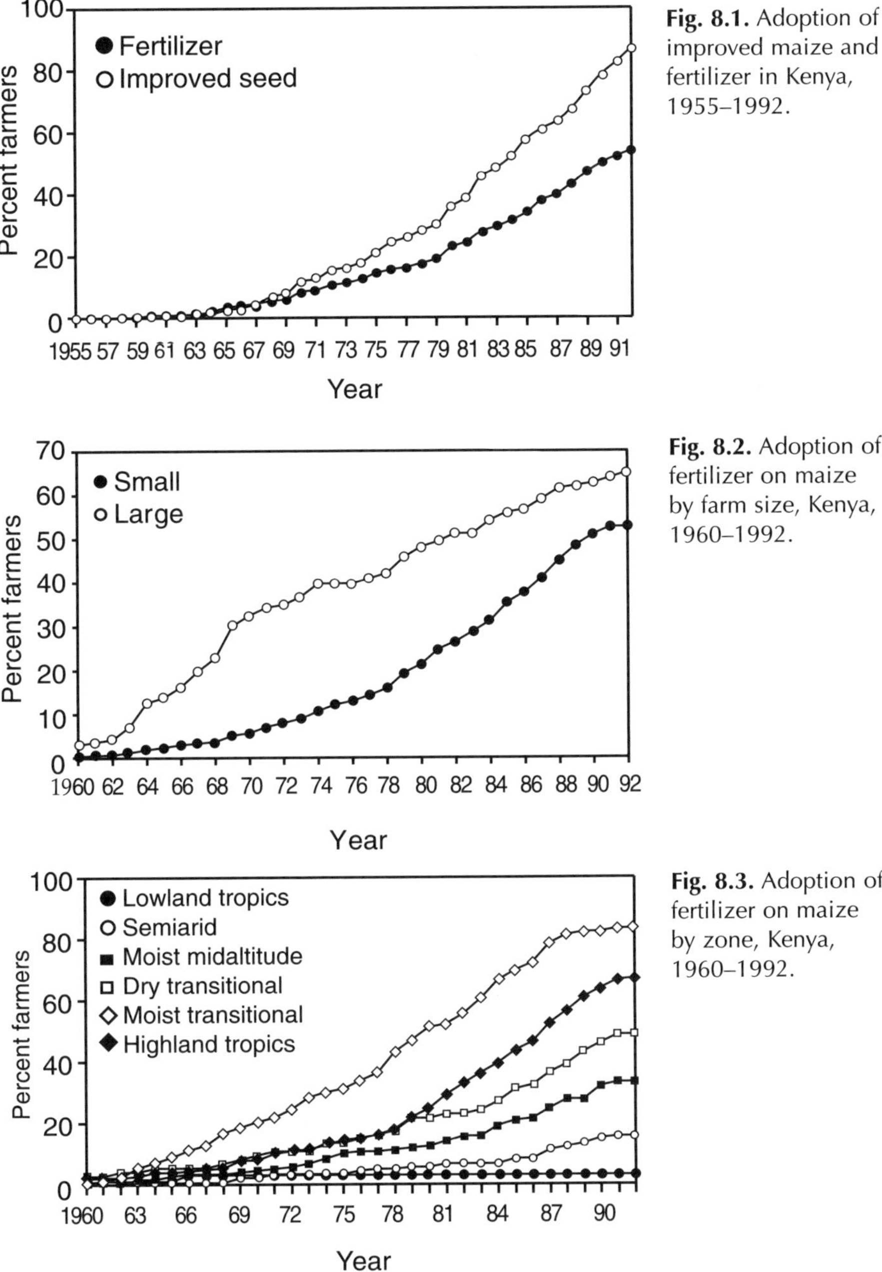

Fig. 8.1. Adoption of improved maize and fertilizer in Kenya, 1955–1992.

Fig. 8.2. Adoption of fertilizer on maize by farm size, Kenya, 1960–1992.

Fig. 8.3. Adoption of fertilizer on maize by zone, Kenya, 1960–1992.

demand to include factors influencing the biological response or potential of fertilizer, scale, risk, credit, awareness, and farmer-specific attributes.

Supply factors

Except for about 6000 t of single superphosphate produced locally each year, Kenya imports all its fertilizer needs, which reached 265,000 t in 1993/94

(Government of Kenya, unpublished report). On average, about 50% of all fertilizer imported into Kenya over the past two decades has been contributed by donors (Murithi and Shiluli 1993, Government of Kenya, unpublished report). Prior to 1989, the government controlled fertilizer marketing through import licensing and price controls. Import licenses were issued to parastatals, such as the Kenya Grain Growers Cooperative Union (KGGCU), Kenya Tea Development Authority (KTDA), and coffee farmers' cooperative unions, and to large estates and private distributors. The KGGCU is the major source of fertilizer distributed to small traders and cooperative societies (Egerton University 1993; Murithi and Shiluli 1993).

The government's control of fertilizer imports and setting of distribution margins and retail price ceilings has been considered a major constraint to the spread of fertilizer use in Kenya (Lele *et al.* 1989; J.H. Allgood *et al.* unpublished report). Domestic price controls were removed in 1989, but the government maintained control over fertilizer imports. These controls, combined with inadequate distribution networks, lack of credit to private suppliers, poor roads and hence high transportation costs, and smaller markets, have led to very low marketing margins and severely limited private traders' competitiveness in fertilizer marketing compared with the well-established KGGCU and other parastatals. Although fertilizer imports are decontrolled now, most of the constraints just listed remain major barriers to the increased efficiency of private fertilizer marketing in Kenya (Egerton University 1993; Murithi and Shiluli 1993; Olouch-Kosura and Chege 1994).

Types and levels of fertilizer used by maize farmers surveyed by the MDBP in the 1992/93 cropping season are summarized in Table 8.2. This table also presents results of a village-level survey of marketing and agricultural support services infrastructure, which was conducted simultaneously at the MDBP survey sites. Significant variation can be observed in the rate of adoption of inorganic fertilizers by agroclimatic zone, with higher adoption in more favorable environments (medium- and high-potential zones). While this pattern reflects the higher agronomic potential of fertilizer use in favorable environments, it also suggests some correlation with the spatial distribution and relative state of development of input delivery and marketing infrastructure.

It is very clear from Table 8.2 that maize farmers in medium- and high-potential zones enjoy much better access to fertilizer supply networks than farmers in low-potential areas. The more favorable zones have a higher percentage of villages that have fertilizer dealers and are in proximity to all-weather roads. On the other hand, the availability of fertilizer to maize farmers in the more marginal lowland tropics and semiarid zone is severely limited by the much longer distances they have to cover, on relatively poorer roads, to buy fertilizer. Poor roads also mean higher transportation costs, lower marketing margins, and ultimately weaker financial incentives for private dealers to deliver fertilizer to remote villages with relatively small markets. Given the bulkiness of fertilizer compared with improved seed, it is clear that the poor input supply and road networks represent a major constraint to the availability of fertilizer and to its increased use in relatively marginal environments. Other studies that have surveyed fertilizer dealers (Murithi and Shiluli 1993;

Table 8.2. Fertilizer use by type and input delivery infrastructure for each agroclimatic zone (1992/93 MDBP Farmer Survey).

	Low-potential zones		Medium-potential zones		High-potential zones		
	Lowland tropics	Semiarid	Moist midaltitude zone	Dry transitional	Moist transitional	Highland tropics	Total
Basal fertilizer (% used, 1992/93)	2	12	24	43	78	59	54
Amount (kg nutrients ha^{-1})	na	15.5	19.3	32.9	40.2	36.6	37.2
Percent diammonium phosphate	100	33	91	59	77	79	73
Percent 20–20 (NPK)	0	50	9	22	10	11	14
Topdressing (% used, 1992/93)	0	3	4	19	29	14	15
Amount (kg nutrients ha^{-1})	0	0.5	0.2	1.8	18.3	5.4	6.8
Percent calcium ammonium nitrate	0	67	50	74	67	79	73
Total kg nutrients ha^{-1}	na	16.0	19.5	34.7	58.5	42.0	44
Animal manure (% used)	22	57	52	89	45	50	48
Percent reported no input supplier in village	80	100	75	50	41	39	56
Distance to nearest market (km)	9	10	18	10	7	10	10
All-weather road to nearest market (% sites)	0	22	12	40	23	44	28

Note: reported amounts of fertilizer applied refer to average for all farmers.

Egerton University 1993) have revealed that, whereas the majority of small farmers prefer 10–25 kg packages of fertilizer, most fertilizer is supplied in 50 kg packages.

Another important feature of fertilizer use is the fact that nearly 90% of the total nutrients applied come from fertilizer types that are relatively low in nitrogen (N) content, such as diammonium phosphate (DAP) and 20:20:0 NPK (Table 8.2). Low-analysis calcium ammonium nitrate (CAN) is the primary source of N for topdressing. The result is a large imbalance in the required nutrients, leading to overuse of phosphorus (P) and underuse of N (see Table 8.5). The major cause of this phenomenon can be attributed to supply, as DAP is the major commercial fertilizer marketed. This fertilizer accounts for more than two-thirds of the fertilizer imported into Kenya (Government of Kenya, unpublished report). The dominance of DAP can also be attributed to its relative abundance in donors' own markets, which makes DAP easier to obtain and supply irrespective of what is needed in recipient countries (Mwangi 1995).[2] Whereas high P application levels are desirable, given the general deficiency in available P in most Kenyan soils, the low levels of N applied are the chief concern in the nutrient imbalance.

Demand factors

As with other production inputs, the demand for fertilizer is derived from its contribution to the value of output. Two elements determine returns to fertilizer use:

1. A measure of the technical relationship between the quantity of output (Y) produced at different levels of fertilizer application (X), keeping other factors (Z) constant. This measure is called the production function:

$$Y = f(X{:}Z). \tag{8.1}$$

2. The unit price of output (P).

Based on the profit maximization assumptions of the theory of the firm, optimal (profit-maximizing) fertilizer levels are reached at the point where the value of additional output units (value of marginal product, VMP) from an extra unit of fertilizer is equal to its cost (price of fertilizer, r):

$$\text{VMP} = P\frac{\partial Y}{\partial X} = r. \tag{8.2}$$

The VMP defines the input demand curve. The point of economic (allocative) efficiency is different from the point of agronomic or production efficiency, where total output is at its maximum, e.g., marginal physical product equals zero (for more details see Jauregui and Sain 1993). According to equation (8.2), the optimality conditions for economic efficiency (profit maximization) require that the marginal product (MP) of the last unit of fertilizer added be equal to what is commonly referred to as the nutrient–grain price ratio:

$$\text{MP} = \frac{\partial Y}{\partial X} = r/P. \tag{8.3}$$

The agronomic response function specified in equation (8.1) exhibits diminishing marginal production, where the gain in maize yield from fertilizer units added at lower levels of soil fertility is higher than yield gains at higher fertility levels. In other words, yield gains from fertilizer use diminish as the concentration of nutrients in the soil increases. Accordingly, response to fertilizer is influenced by the initial level of soil nutrients. In dryland agriculture, soil moisture is determined by physical soil characteristics (water holding capacity), together with climatic factors such as rainfall. In turn, soil moisture influences the yield response to fertilizer. Also, management factors such as timing and method of fertilizer application, weeding, the right type of fertilizer to provide the optimal nutrient balance, use of responsive seed types, and the type of cropping system (sequence and mix), among other factors, significantly influence crop response to fertilization.

Measuring Maize Response to Fertilizer

Records from five years of fertilizer experiments by FURP during 1985–1989 are used to estimate fertilizer response parameters for maize. The FURP experiments were conducted at 70 sites over a wide range of climates, soil types, and systems of maize farming (KARI 1990). In these experiments, various combinations of N and P levels were tested under sole and mixed maize cropping and various seed types. The FURP sites were geo-referenced and the resulting digitized map was overlaid on the MDBP zonal map. This procedure made it possible to locate and classify FURP sites by agroclimate and to integrate all experimental records and results with the spatial and survey components of KARI's maize database. The combined database was employed to analyze maize yield response to fertilization.

Two experiments were conducted under the FURP, characterized as Type 1 and Type 2 experiments (see Appendix F for details). The FURP data set generated by the Type 1 experiments was used to analyze maize yield response to N and P fertilizers. Experimental factors tested in the Type 1 trials were:

1. Levels of N and P.
2. Mode of production, distinguishing between pure stand and intercropped maize.
3. Soil type. Soils of the FURP experimental sites belong to ten soil classes (KARI 1990). Sites were grouped into four soil categories as follows:[3]

- Low fertility. This group includes soil classes with high acidity and low availability of nutrients (e.g., acrisols and ferrasols) and soils with poor physical properties, such as low moisture retention capacity and high leaching (e.g., sandy soils).
- Medium fertility. Soils of moderate fertility and organic matter content, such as planosols, luvisols, cambisols, and gleysols.
- High fertility. Soils of high nutrient availability and organic matter content, such as nitisols, vertisols, and phaesoems.
- Andosols have high organic matter but limited P availability because

of phosphorus fixation, and thus they are placed in a separate category.

4. Maize cultivar. One of the deficiencies of the FURP fertilizer experiments is the fact that they only tested improved maize cultivars. Therefore, the experiments did not generate comparative information on the responsiveness of local varieties to fertilization. Six cultivar groups are defined to control for the variety effect:

- The Coast Composite, recommended for the lowland tropics.
- Pwani hybrid, recommended for the lowland tropics.
- Dryland composites Katumani and Makueni, recommended for the semiarid zone.
- Medium maturity hybrids (500 series), recommended for the midaltitude zones.
- Late-maturity varietal hybrids H622 and H614.
- Late-maturity conventional hybrids H625 and H6262.

The first two cultivars were tested only in the lowland tropics, the zone for which they are recommended. Dryland composites were tested against the 500 series in the semiarid and dry transitional zones. The 500-series hybrids were tested against both varietal and conventional 600-series hybrids in the moist midaltitude zone (western Kenya) and against conventional 600-series hybrids in the high-potential zones (the moist transitional zone and the highland tropics).

Results of an analysis of variance of the effect of factors in the Type 1 experiments on maize yield are summarized in Table 8.3. The effects of variations in all factors – climate, soil, cropping system, cultivar, and levels of N (NTR) and P (PHS) – on maize yield were statistically very significant. Similarly, interaction terms between NTR, PHS, and soil type were highly significant in influencing maize yield response.

A transcendental function was specified to measure maize yield response to N and P fertilizers, controlling for the effects of the regressors listed above. Interactions between N and P and between the two nutrients and the various soil types were also tested. The transcendental function is described in Appendix G. Estimates of the parameters of the maize yield–fertilizer response function are given in Table 8.4. Error statistics (R^{-2}, F- and t-ratios) indicate that all equations provided reasonable explanations of the variation in maize yield response to fertilizer. According to the transcendental specification, marginal products of N (NTR) and P (PHS) are not constants but vary with nutrient application levels and other factors in the interaction terms (Table 8.4). Therefore, the marginal gain from N and P can be determined only after the arguments of the MP function are evaluated (see the following section and Table 8.5).

However, Table 8.4 reveals important patterns of maize yield response to fertilizer. First, estimates of the parameter measuring the interaction between levels of N and P, although not always significant, are consistently positive, indicating the complementarity and mutually enhancing effects of both

Table 8.3. Analysis of variance results of maize yield response to nitrogen and phosphorus in Type 1 experiment (FURP 1985–1989).

Source of variation	*F*-ratio
Covariates:	664.4***
Agroclimatic zones	45.2***
Cropping pattern	13.4***
Cultivar	472.4***
Main effects:	154.1***
Nitrogen	200.3***
Phosphorus	41.9***
Soil type	219.7***
Two-way interactions:	3.5***
Nitrogen–phosphorus	4.1***
Nitrogen–soil type	3.8***
Phosphorus–soil type	2.4***
Explained	89.1***
Number of cases	9196

Note: *** indicates statistical significance at 99% level.

nutrients. Second, except in the lowland tropics, it is clear that maize is more responsive to fertilization when it is planted in a pure stand than when it is intercropped (as shown by the highly significant mode of production effect). The relatively low fertility of soils tested in the lowland tropics may be the reason for the higher maize response under intercropping with leguminous crops. This result may also be caused by a varietal effect indicating that the cultivars tested in the lowland tropics (Coast Composite and Pwani hybrid) perform better under intercropping than in pure stand.

Third, soil type effects are statistically significant, which suggests that a lower response obtains in soils of low fertility. This finding is especially true for marginal environments (the lowland tropics and semiarid zone). In these environments, most of the experimental sites had sandy soils with low humus content – hence low inherent fertility – and poor physical properties (low moisture retention and high leaching) (KARI 1994a, 1994b). Other things being equal, a higher response to fertilization is generally expected from soils with low nutrient availability than from soils with a high nutrient base, but several other factors work to limit fertilizer response in soils of low inherent fertility and poor physical properties. In marginal environments, where leaching of added nutrients is quicker and moisture holding capacity is poorer, sandy soils dominate the group of low fertility soils, which may explain their lower response to fertilization, especially in zones of high moisture stress. This

Table 8.4. Parameter estimates of the transcendental maize yield response to nitrogen and phosphorus function in Type 1 experiment (FURP 1985–1989).

	Lowland tropics	Semiarid zone	Moist midaltitude zone	Dry transitional zone	Moist transitional zone	Highland tropics
Constant term	7.13***	7.09***	7.51***	6.6***	7.46***	7.84***
Nitrogen log term (ln N)	0.036***	0.039***	0.017***	0.025*	0.004	0.007**
Nitrogen level (N)	0.005**	–0.002	–0.0001	–0.0001	0.005***	0.002**
Phosphorus log term (ln F)	0.009	0.029***	0.021***	0.023*	0.022***	0.012***
Phosphorus level (F)	–0.0001	–0.002	–0.0001	–0.0001	–0.001	–0.0001
N–P interaction term (N×F)	0.0001*	0.0001***	0.0001	0.0001	0.0001	0.0001
Production mode (pure stand)	–0.218***	0.152***	0.033***	0.407***	0.06***	0.072***
Soils						
Low fertility (LF)	–0.175**	–0.377***	–0.038	–	–0.041	0.523***
Medium fertility (MF)	–	–	0.262***	0.099	0.979***	0.035
High fertility (HF)	–	0.377	–0.541***	–0.099	0.296***	–0.595***
LF interaction with N	–0.001	0.003***	0.0001	–	–0.067	–0.0001
LF interaction with P	0.001	0.002**	0.001***	–	–0.003***	–0.001
MF interaction with N	–	–	0.002*	–0.003**	0.003**	–0.0001
MF interaction with P	–	–	–0.0001	–0.003**	0.002*	–0.003***
HF interaction with N	–	–	–0.003	–	–0.004*	0.004***
HF interaction with P	–	–	–0.002	–	–0.005*	0.002***
Varieties						
Coast Composite	0.041*	–	–	–0.114*	–	–
Dryland composites	–	–0.344***	–	–0.114	–	–
H500 hybrids	–	0.344	–0.09***	–	–0.191***	–0.364
Varietal 600 hybrids	–	–	0.123***	–	–	–
Conventional 600 hybrids		–	–0.033	0.191	0.364	
R^2	0.19	0.29	0.16	0.16	0.19	0.34
F–ratio	25.6***	47.5***	18.1***	11.35***	40.6***	59.5***
Number of cases	1085	1216	1660	605	2778	1846

Note: *, **, *** indicate statistical significance at 90, 95 and 99% level, respectively.

is confirmed by the significantly higher yield response base in relatively favorable environments (up to 0.523 kg grain yield difference in the highland tropics) where moisture stress is less limiting and inherent soil fertility relatively higher. In all zones a greater response is achieved consistently in soils of medium fertility compared with soils of high fertility, which once again is consistent with the importance of inherent fertility levels and physical soil properties for a higher response to fertilizer.

Interactions between soil and fertilizer types (different slopes or MPs) show mixed results and require more careful interpretation. The grouping of soil types used in this study is not adequate for analyzing such interactions, as valuable information on initial soil nutrient levels is lost by this very broad grouping. More refined grouping by nutrient levels is certainly needed if we are to gain a better understanding of these interactions. Nevertheless, the results show a positive interaction between P levels and soils with P fixation problems,[4] i.e., higher marginal MP for PHS.

Finally, an unexpectedly higher yield response base (intercept) is achieved with composite maize cultivars than with hybrids in the lowland tropics (Coast Composite vs. Pwani hybrid) and the dry transitional zone (dryland composites vs. H500 hybrids). However, the H500 hybrids showed higher responsiveness to fertilization than dryland composites in the semiarid zone. These results provide no conclusive evidence for the superiority of composites over hybrids in more marginal environments and under drought stress. The important interaction between seed type and agroclimate is further confirmed by the results in the relatively moist environments. Late maturity varietal hybrids (H622 and H614) outperform medium maturity hybrids (H500) and late maturity conventional hybrids (H625 and H626) in the relatively warmer moist midaltitude zone. However, in the cooler environments of the moist transitional zone and highland tropics, late maturity conventional hybrids (H625 and H626) show a much higher response to fertilization than the medium maturity H500 hybrids. These results show clearly the importance of matching suitable germplasm with agroclimate for highest response levels.

Agronomic and allocative (economic) efficiency levels of N and P are computed and compared with average levels applied by farmers on maize in the 1992/93 season (Table 8.5). Results are summarized by zone to compare optimal rates and farmers' average application rates (obtained from the MDBP survey data, in which sites were sampled by agroclimate rather than soil type). Accordingly, the maize-yield response function was re-estimated by zone to provide an average response measure for the zone (over soil type, seed type, and mode of production). The results indicate that economically optimum levels calculated for the 1992/93 season are much lower than the agronomic efficiency levels because of the relatively high nutrient–grain price ratio.

It is important to note that economic efficiency levels are derived with the assumption that no extra labor or operating costs (interest) are associated with fertilizer use. Given the fact that fertilizer usually constitutes a significant production cost item to finance, especially among small-scale farmers, a value–cost (VMP) of more than twice the nutrient-grain price ratio is generally required for economic efficiency (Jauregui and Sain 1993; Heisey and Mwangi

Table 8.5. Optimal levels of nitrogen and phosphorus fertilizers on maize by zone and soil type (FURP Type 1 experiment, 1985–1989).

	Lowland tropics	Semiarid zone	Moist midaltitude zone	Dry transitional zone	Moist transitional zone	Highland tropics
Nitrogen (kg ha^{-1})						
Agronomic efficiency[a]	154	105	73	77	166	121
Economic optimum for 1992/93[b]	98	70	55	40	88	85
Average amount used in 1992/93	na	5	5.6	10.8	23.1	15.7
Extraction rate, 1992/93[c]	17.7	13.4	18.7	15.7	35.9	37.8
Balance, 1992/93[d]	na	−8.4	−13.1	−4.9	−12.8	−22.1
Price ratio, 1992/93[e]	5.8	5.8	5.8	5.8	5.8	5.8
Price ratio, 1994/95[f]	7.3	7.3	7.3	7.3	7.3	7.3
Phosphorus (kg ha^{-1})						
Agronomic efficiency[a]	243	161	96	98	61	217
Economic optimum for 1992/93[b]	92	73	42	36	59	75
Average amount used in 1992/93	na	11	13.9	23.9	35.4	26.3
Extraction rate, 1992/93[c]	8.2	6.2	8.7	7.3	16.6	17.5
Balance, 1992/93[d]	na	4.8	5.2	16.6	18.8	8.8
Price ratio in 1992/93[e]	3.5	3.5	3.5	3.5	3.5	3.5
Price ratio in 1994/95[f]	5.5	5.5	5.5	5.5	5.5	5.5
Average maize yield, 1992/93 (kg ha^{-1})	1360	1030	1440	1210	2760	2910

Note: na = data not available.

[a] Agronomic efficiency levels are calculated as input levels that maximize physical output (e.g., MP=0).

[b] Economic optimum levels are calculated as input levels at which MP is equated to the nutrient price ratio (equation 8.3).

[c] Based on a rate of extraction (in grain and crop residue forms) of 15 kg nitrogen and 6 kg phosphorus per 1000 kg of harvested product (i.e., yield). Adapted from FAO (1987) and Ransom *et al.* (1993).

[d] Represents the difference between total nutrients extracted from the soil in harvested product and total supplied from fertilizer.

[e] Government of Kenya (1993). The same average price ratio was used for all zones, assuming that variation in input prices (numerator) caused by location cancel out with similar variations in output prices (denominator).

[f] Government of Kenya (unpublished report).

1997). If labor and financial costs are considered, economically optimum levels of N and P for the 1992/93 season will be even lower than the agronomic optimum. It is also very clear that the nutrient–grain price ratio has deteriorated for maize against fertilizer use between 1992 and 1994 (Table 8.5). However, farmers applied much lower quantities of N and P than economically optimum rates in 1992/93, especially in the marginal zones.

When nutrient extraction rates are approximated based on the 1992/93 maize-yield levels and subtracted from the average levels of N and P applied, significant mining of N (a negative balance) is observed. As discussed earlier, the use of fertilizers with low N content, such as DAP, is a major cause of the imbalance between N and P levels and, consequently, the rate at which the two nutrients are mined from the soil. The nutrient balance calculations reported in Table 8.5 must be considered very crude approximations, because they do not take into account other sources of nutrient supply and extraction (i.e., use of animal manure, atmospheric deposition, biological N fixation, losses due to erosion, leaching, and so on). Nevertheless, our results are similar to estimates of soil nutrient mining rates in Kenya and Africa in general that have been generated by specialized studies based on more comprehensive nutrient balance modeling (FAO 1987; Van der Pol 1992; Ransom *et al.* 1993; Smaling *et al.* 1993a; Smaling *et al.* 1993b; Stoorvogel *et al.* 1993).

These results confirm that to increase the marginal gains from inorganic fertilizers it is very important to target fertilizer recommendations effectively, by soil type, cultivar, agroclimate, and mode of production. The economic return to fertilizer use is also a crucial factor influencing demand for fertilizer. Therefore a key strategy for promoting the use of fertilizers is to develop economic policies and institutional measures that would improve the nutrient–grain price ratio for higher returns to farmers. Such policies and measures would thus contribute to higher food production and to protecting soils from excessive nutrient mining. Examples of such policies and strategies include infrastructural development for lower transport and transaction costs, an effective credit system, and the importation and distribution of the right type of fertilizer (fertilizer with a high N content).

In addition to the determinants of fertilizer profitability we have just discussed, other factors influence farmers' demand for fertilizer. Two very important demand factors are farmers' awareness of fertilizer options (either as a result of extension services or other sources of information) and farmers' financial ability to purchase fertilizer (affordability). A number of factors affect affordability, such as farmers' access to credit, cash income, and the availability of self-financing, as is the case with large-scale farmers. The effects of these factors on fertilizer use by maize farmers are summarized in Tables 8.6 and 8.7.

More farmers adopted fertilizer in the medium- and high-potential zones, because they had relatively better access to extension and credit services than farmers in the low-potential zones. Almost all large-scale farmers, compared with about half of small-scale farmers, adopted fertilizer (Table 8.7). Although large-scale farmers had better access to extension and credit services than small-scale farmers, the percentage of farmers who applied fertilizer without credit is the same in the two groups (the percentage who used fertilizer minus

Table 8.6. Access to extension and credit for fertilizer by zone (percentage of farmers).

Attribute	Lowland tropics	Semiarid zone	Moist midaltitude zone	Dry transitional zone	Moist transitional zone	Highland tropics	Average
Used fertilizer in 1992/93	2	12	24	43	78	59	50
Received advice on fertilizer	36	47	56	55	74	57	59
Used credit to purchase fertilizer	0	3	1	1	6	7	5
Obtained credit from formal sources[a]	0	0	100	0	67	90	80

Source: MDBP Farmer Survey, 1992/93 season.

[a] Percentage of credit users. The Agricultural Finance Corporation, Banks, and KGGCU were reported as the formal sources of credit. Other sources of credit reported include NGOs.

Table 8.7. Access to extension and credit for fertilizer by farm size, maize seed type, and tea/coffee farming (percentage of farmers).

	Farm size		Maize seed type		Cash crop	
Attribute	Small (<8 ha)	Large (>8 ha)	Hybrid	Other	Tea/ coffee	No tea/coffee
Used fertilizer, 1992/93	47	97	68	8	83	44
Received advice on fertilizer	55	80	73	54	64	40
Used credit to purchase fertilizer	2	53	6	0	12	3
Obtained credit from formal sources[a]	70	92	83	0	33	90

Source: MDBP farmer survey, 1992/93 season.

[a] Percentage of credit users. The Agricultural Finance Corporation, Banks, and KGGCU were reported as the formal sources of credit. Other sources of credit reported include NGOs.

the percentage who received credit). This result indicates that the scale effect works only indirectly through the ability to acquire credit. On the other hand, the effect of seed type was stronger than the effect of credit availability. The percentage of farmers who used fertilizer on hybrids without credit was much higher than the percentage of fertilizer users among farmers planting local and composite varieties.

Fig. 8.4 compares fertilizer adoption on hybrids versus other maize cultivars over the past three decades. It is important to note that hybrids are mainly used in the medium- and high-potential zones (Chapter 7), where the climate favors a higher response to fertilizer. This result has several implications. First, it suggests that profitability (higher yield gains) is more important than the availability of credit for promoting the demand for fertilizer among maize farmers. Second, it indicates the crucial role that increased adoption of hybrid seed can play in encouraging fertilizer use, especially among smallholders in the midaltitude zones where adoption is still low (Chapter 7). Finally, it serves

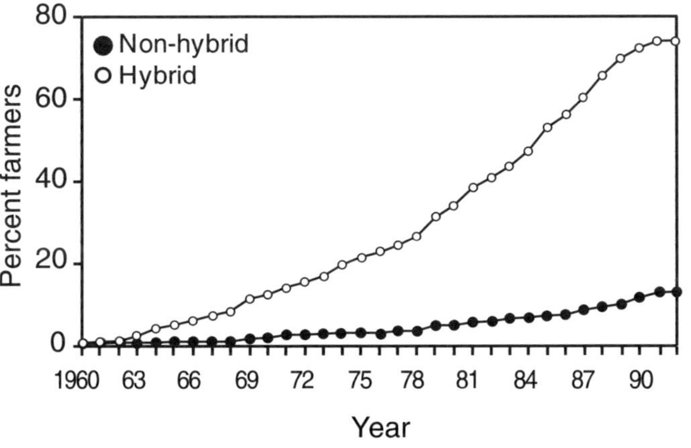

Fig. 8.4. Adoption of fertilizer on maize by cultivar, Kenya, 1960–1992.

to emphasize the importance of maize breeding and crop management research for the development and proper targeting of appropriate germplasm and fertility management technologies that can improve the returns to fertilizer use on maize.

Similarly, the availability of fertilizer at subsidized prices through tea or coffee farming[5] or access to cash income had a stronger impact on fertilizer use than the availability of formal credit to maize farmers (Table 8.7). However, one should not forget that on average very few maize farmers (5%) received credit for fertilizer (Table 8.6).

Empirical Fertilizer Adoption Models

Formal empirical models were estimated to measure and test hypotheses about the correlation between fertilizer adoption and the set of supply and demand factors discussed above. A logistic regression analysis of the probability of fertilizer adoption was performed on the MDBP survey data (Table 8.8). The effect of the same set of regressors on the intensity of fertilizer use was also analyzed using the tobit model[6] (Table 8.9). The sex, age, and level of education of the head of the farming family were added to the set of supply and demand factors reviewed above.

Results of the empirical analysis confirmed the significance of the supply and demand forces discussed previously as determinants of fertilizer use. Sup-

Table 8.8. Results of the logit analysis of fertilizer adoption.

Dependent variables:	Adoption (yes/no)	Parameter estimates	
Model χ^2	409.4***		
Goodness of fit	847.7		
Valid cases	902	β	Exp(Z)
Exogenous variables:			
Used hybrid seed		1.44***	4.20
Received credit		1.02**	2.76
Male		−0.07***	0.94
Area sown to maize (ha)		0.08***	1.08
Received education		0.05	1.05
Age (years)		−0.01**	0.99
Road to primary market (all-season)		0.22**	1.24
Distance to primary market (km)		−0.05***	0.98
Received extension advice		0.36***	1.43
Home plot (*shamba*)		0.09	1.09
Monocropped maize		−0.11	0.9
Tea or coffee farmer		0.75***	2.12
Constant		1.14*	–

Note: *, **, *** indicate statistical significance at 90%, 95%, and 99% levels, respectively.

Table 8.9. Results of the tobit analysis of fertilizer use on maize.

Dependent variable:	Amount of fertilizer (kg ha^{-1})		
R^2	0.36		
F-ratio	86.8***	Parameter	
Number of cases	1406	estimates	*t*-value
Exogenous variables:			
Used hybrid seed		3.45	20.6***
Received credit		1.40	3.8***
Male		−0.20	−1.3
Area sown to maize (ha)		0.51	3.8***
Road to primary market (all-season)		0.83	4.6***
Distance to primary market (km)		−0.44	−2.6***
Received extension advice		0.79	4.9***
Monocropped maize		−0.08	−0.4
Tea or coffee farmer		1.38	5.5***
Constant		−0.86	−1.7*

Note: *, **, *** indicate statistical significance at 90%, 95%, and 99% levels, respectively. Dependent variable and continous regressors are in natural logarithm.

ply forces were modeled through the effects of road infrastructure variables. Both factors were very powerful in explaining variations in fertilizer adoption, as seen from the significance of their respective residual statistics (Tables 8.8 and 8.9). Returns to fertilizer, as approximated by the use of hybrids versus other cultivars, proved to be the most important determinant of fertilizer use among supply and demand factors, both in terms of its influence on the odds of adoption and the intensity of use (magnitude of the regression coefficient in Table 8.9). Access to credit and cash income or subsidized (low-cost) fertilizer through tea or coffee farming came second in importance. These results are in accordance with those reported for other parts of Africa (Mwangi 1995).

However, contrary to the results of the FURP experiments, the odds of fertilizer adoption were lower for farmers planting maize in pure stand. This result may be attributed to the very high proportion of farmers in the MDBP sample who intercrop maize (Chapter 4). Farmer characteristics were important, for example for the probability of adoption, but insignificant determinants of intensity of use.

Fertilizer Use and the Maize Yield Gap

The preceding analysis showed that maize farmers apply much less than the optimal level of fertilizer (Table 8.5). Here we compare yield levels achieved by farmers under current practices with economically optimal yield levels derived from the fertilizer response analysis of the FURP data. Economically

Table 8.10. Gap between Kenyan farmers' maize yields and economically optimum yield levels in the 1992/93 cropping season, by zone.

	Lowland tropics	Semiarid zone	Moist midaltitude zone	Dry transitional zone	Moist transitional zone	Highland tropics
Ratio of phosphorus to nitrogen						
Economic optimum in 1992/93[a]	0.9	1.1	0.8	0.9	0.7	0.9
Levels used in 1992/93	na	2.2	2.5	2.2	1.5	1.7
Yield (kg ha^{-1})						
At economic optimum levels[a]	2020	1560	2570	1810	3590	4170
Farmers' average[b]	1360	1030	1440	1210	2760	2910
Yield gap (ratio of current to optimal)	0.67	0.66	0.56	0.67	0.77	0.70
Total area under maize (000 ha)	33	118	118	37	424	307
Potential production increase (000 t)[c]	21.8	62.54	133.34	22.20	351.92	386.82
Percentage of total gain in production	2.2	6.4	13.6	2.3	36.0	39.5
Percentage of farmers used improved seed in 1992/93	16	33	41	21	85	86

[a] Based on the FURP experimental records for 1985–1989 (KARI 1990).
[b] Average yields achieved by MDBP sample farmers during the survey year (1992/93).
[c] Calculated as the yield gap (economic optimum minus farmers' average yield) multiplied by area under maize in the zone.

optimal levels of N and P calculated in Table 8.5 are used to compute optimal yield levels at the 1992/93 nutrient–grain price ratio (Table 8.10). The gap in maize yield, defined as the ratio of farmers' yield to the economically optimum yield level, is derived. The largest gap is observed in the moist midaltitude zone (1−0.56), where the balance between N and P is the most skewed, at more than three times the desired ratio. In general, however, more than twice the optimal nutrient balance is applied in all zones, indicating high overuse of P and underuse of N. As mentioned earlier, the major cause of this severe imbalance in the required composition of nutrients is that fertilizer types that are low in N (such as DAP) predominate among commercial fertilizers traded in Kenya.

On average, there is a gap of about 30% between farmers' yield levels and the economically optimal yield. It is important to note that the FURP yield levels are much lower (about half) than yields recorded on the experiment station (KARI 1993). This difference can be attributed to the fact that FURP trials were conducted practically under farmers' conditions, and hence they can be considered to represent a realistic potential. The gain from achieving the economically optimal yield level would amount to about one million additional tonnes of maize. More than 75% of this potential increase in production would come from the moist transitional zone and the highland tropics (Table 8.10), because the high intensity of maize cultivation (area under maize) and higher yield levels in these areas result in a higher production base.

It is very clear from Table 8.10 that nonoptimal soil fertility management is the major source of the yield gap in the moist transitional zone and highlands, because almost all farmers in these zones have adopted improved seed. The potential for improving productivity by using the right levels and composition of nutrients is tremendous. Another significant source of potential productivity gains can be exploited in the moist midaltitude zone, where adoption of improved seed is still very low. Optimal fertilizer levels and nutrient balance, in combination with the increased use of improved maize seed, could bring about a considerable productivity increase.

Conclusion: Implications for Policy and Research

Kenyan farmers apply lower rates of inorganic fertilizers on their maize crop than is considered economically optimal. More importantly, a serious imbalance in the ratio of applied N to P causes the rate of N extraction from the soil to exceed the rate of supply, leading to mining of soil N in maize farming. The gap between farmers' yields under current practices and the yields that could be attained if optimal levels of N and P were applied is considerable. About one million more tonnes of maize could be added to current domestic production (an increase of one-third) if farmers improved their soil fertility management practices. More than 75% of that gain would be realized in the moist transitional zone and highland tropics, where almost all large- as well as small-scale farmers have adopted improved seed.

We have also presented empirical evidence related to the two factors that are crucial for increasing farmers' demand for fertilizer: lower nutrient-grain price ratios and high yield gains from fertilizer use. Among the supply side forces that influence the price of fertilizer are transportation and transaction costs. The poor road infrastructure and the long distances farmers have to travel to purchase fertilizer are the major cause of high fertilizer costs, especially for smallholders in more marginal and remote areas. Transportation and transaction costs also lower the marketing margins and hence form a major disincentive to private traders to enter and engage in fertilizer delivery, especially to small markets in remote areas. High transportation costs erode the profitability of fertilizer and reduce the incentive for farmers' demand. Evidence for the relatively high transportation and transaction costs associated with maize marketing in Kenya has been provided by Donovan (1995), who showed that in 1992 marketing costs per t km^{-1} of maize imported from the US into western Kenya were almost five times as great as they were in the US, despite the fact that the distance traveled by maize from farms to the US Gulf ports was about 1.5 times the distance between Mombasa and western Kenya.

The other major supply factor is the dominance of low N fertilizers among the commercial products traded in Kenya. Promotion of high N products such as urea will reduce the relative cost of the N component, lower the nutrient–grain price ratio, and improve the returns to fertilizer use for maize farmers. In addition, public investment in rural roads, removal of restrictions on fertilizer imports, and the provision of credit and tax incentives to private traders to invest in private storage and distribution facilities would have a significant impact on the diffusion of fertilizer. The marketing of fertilizer in smaller packages (10–25 kg), which are easier to transport and more economic, would also help increase fertilizer use, especially among small-scale farmers in remote areas.

One more potential means of increasing the profitability of fertilizer use would be to achieve higher yield gains (biological response) from fertilizer application. Our study has shown high fertilizer use to be associated with higher adoption of hybrid maize, mainly because hybrids respond well to fertilization, especially in the high-potential zones. Although maize hybrids are also recommended for the medium-potential zones, their use is still very low, and greater adoption of hybrids would contribute to higher use of fertilizers in these zones. However, the development of better adapted hybrids that are acceptable to farmers in these zones remains a challenge for maize research (Chapter 7).

This study has also confirmed the importance of soil type and level of soil nutrients in determining optimal fertilizer rates. Coordinated crop management research and extension efforts to disseminate fertilizer recommendations conditioned by soil type and nutrient analysis are crucial for improving the agronomic (yield) response (and hence the returns) to fertilizer use. Alternative low-cost sources of N, such as green and animal manures for efficient nutrient recycling, and methods of soil conservation, will reduce soil nutrient depletion and enhance maize productivity in the long run (Qureshi 1987). The improvement of soil fertility continues to be the real challenge for crop management research in the low-input, maize-based farming systems of smallholders in Kenya and elsewhere sub-Saharan Africa.

References

CIMMYT (1992) *CIMMYT World Maize Facts and Trends*. International Maize and Wheat Improvement Center (CIMMYT), Mexico City.

Donovan, W.G. (1995) The role of inputs and marketing systems in the modernization of agriculture. Paper presented at the SG-2000 Workshop, 'Developing African Agriculture: Achieving Greater Impact from Research Investments,' 26–30 September, Addis Ababa, Ethiopia.

Egerton University (1993) A review of Kenya's fertilizer marketing reform. PAM, Njoro, Kenya. Draft report.

FAO (Food and Agriculture Organization) (1987) *Fertilizer Strategies*. FAO, Rome, Italy.

Gerhart, J. (1975) *The Diffusion of Hybrid Maize in Western Kenya*. Research Report for CIMMYT. International Maize and Wheat Improvement Center (CIMMYT), Mexico City.

Government of Kenya (1993) *Statistical Abstracts*. Central Bureau of Statistics. Ministry of Planning and National Development. Government Printer, Nairobi.

Heisey, P.W., and Mwangi, W. (1997) Fertilizer use and maize production in sub-Saharan Africa: Use and policy options. In: Byerlee, D., and Eicher, C.K. (eds) *Africa's Emerging Maize Revolution*. Lynne Rienner, Boulder, Colorado, pp. 123–211.

Jauregui, M., and Sain, G. (1993) *Continuous Economic Analysis of Crop Response to Fertilizer in On-farm Research*. CIMMYT Economics Paper No. 3. International Maize and Wheat Improvement Center (CIMMYT), Mexico City.

KARI (Kenya Agricultural Research Institute) (1990) *The Fertilizer Use and Recommendations Project (FURP): Manual 1*. National Agricultural Research Laboratories (NARL), KARI, Nairobi, Kenya.

KARI (Kenya Agricultural Research Institute) (1993) *Improve and Sustain Maize Production through Adoption of Known Technologies*. Information Bulletin No. 7. KARI, Nairobi, Kenya.

KARI (Kenya Agricultural Research Institute) (1994a) *Coastal Districts (Kilifi/Kwale/Lamu)*. Vol. 1 of *Fertilizer Use and Recommendations*. Fertilizer Use and Recommendations Project (FURP), National Agricultural Research Laboratories (NARL), and KARI, Nairobi, Kenya.

KARI (Kenya Agricultural Research Institute) (1994b) *Machakos/Makueni Districts*. Vol. 23 of *Fertilizer Use and Recommendations*. Fertilizer Use and Recommendations Project (FURP), National Agricultural Research Laboratories (NARL), KARI, Nairobi, Kenya.

Keating, B.A., McCown, R.L., and Wafula, B.M. (1993) Adjustment of nitrogen inputs in response to a seasonal forecast in a region of high climatic risk. In: Penning de Vries, F.W. *et al.* (eds) *Systems Approaches for Agricultural Development*. Kluwer Academic Publishers, Dordrecht, The Netherlands, pp. 233–252.

Kimuyu, P., Jama, M., and Muturi, W. (1991) Determinants of fertilizer use on smallholder coffee and maize in Muranga District, Kenya. *Eastern Africa Economic Review* 7(1), 1–11.

Larson, A.B. 1993. *Fertilizers to Support Agricultural Development in Sub-Saharan Africa: What Is Needed and Why*. Center for Economic Policy Studies Discussion Paper No. 13. Winrock International, Rosslyn, Virginia.

Lele, U., Christiansen, R.E., and Kadiresan, K. (1989) *Fertilizer Policy in Africa: Lessons from Development Programs and Adjustment Lending, 1970–87*. MADIA Discussion Paper 5. World Bank, Washington, DC.

Murithi, F., and Shululi, M. (1993) Effects of the liberalization of fertilizer markets on the distribution and use of fertilizer on food crop production: A study on Embu and Meru Districts of Kenya. In: Mwangi, W,. Rohrbach, D., and Heisey P.W. (eds) *Cereal Grain Policy Analysis in the National Agricultural Research Systems of Eastern and Southern Africa.* International Maize and Wheat Improvement Center (CIMMYT), Southern African Development Community (SADC), and the International Center for Research in the Semi-Arid Tropics (ICRISAT), Addis Ababa, Ethiopia.

Mwangi, W. (1995) Low use of fertilizers and low productivity in sub-Saharan Africa. Paper presented at the IFPRI/FAO workshop, 'Plant Nutrition Management, Food Security, and Sustainable Agriculture and Poverty Alleviation in Developing Countries,' 16–17 May, Viterbo, Italy.

Nadar, H., and Faught, W. (1984) Effect of legumes on yield of associated and subsequent maize in intercropping and rotation systems. *East African Agriculture. and Forestry Journal* 44, 127–136.

Oluoch-Kosura, W., and Chege, D. (1994) Evolution of fertilizer policy and marketing in Kenya: The search for effective fertilizer distribution and use. Paper presented at the conference, 'Agricultural Policies and Food Security in Eastern Africa,' Department of Agricultural Economics, University of Nairobi, Kenya.

Qureshi, J. (1987) The cumulative effects of N–P fertilizers, manure and crop residues on maize grain yields, leaf nutrient contents and some soil chemical properties at Kabete. Paper presented at the National Maize Agronomy Workshop, Nairobi, Kenya.

Ransom, J., Ojiem, J., and Kanampiu, F. (1993) Nutrient flux between components of maize–coffee–livestock systems in central Kenya. Paper presented at the international conference, 'Livestock and Sustainable Nutrient Cycling in Mixed Farming Systems of Sub-Saharan Africa,' Addis Ababa, Ethiopia.

Smaling, E., Stoorvogel, J., and Windemijer, P. (1993a) Calculating soil nutrient balances in Africa at different scale: District level. Cited in E. Smaling, An agro-ecological framework for integrated nutrient management, with special reference to Kenya. Unpublished PhD thesis, Agricultural University, Wageningen, The Netherlands.

Smaling, E., Nandwa, S., Prestele, H., Roetter, R., and Muchena, F. (1993b) Yield response of maize to fertilizer and manure under different agro-ecological conditions in Kenya. Cited in E. Smaling, An agro-ecological framework for integrated nutrient management, with special reference to Kenya. Unpublished PhD thesis, Agricultural University, Wageningen, The Netherlands.

Stoorvogel, J., Smaling, E., and Janssen, B. 1993. Calculating soil nutrient balances in Africa at different scales: Supra-national scale. Cited in E. Smaling, An agro-ecological framework for integrated nutrient management, with special reference to Kenya. Unpublished PhD thesis, Agricultural University, Wageningen, The Netherlands.

Van der Pol, F. 1992. *Soil Mining: An Unseen Contributor to Farm Income in Southern Mali.* Bulletin 325. Royal Tropical Institute, Amsterdam, The Netherlands.

Endnotes

1 To allow simple comparisons, the final six maize-specific agroclimatic zones of the MDBP (Plate 4) are grouped into high-, medium-, and low-potential zones (the zones listed in Table 8.1). The highland tropics and moist transitional zone constitute the high-potential zone; the dry transitional and moist midaltitude zones form

the medium-potential zone; and the semiarid zone and lowland tropics form the low-potential zone.

2 The average ratio of N, P, and potassium (K) nutrients applied to Kenya's soils between 1984/85 and 1993/94 was 1:0.95:0.12. The nutrients composition ratio on maize for the same period was more skewed at an average of 1:1.4:0 (Government of Kenya, unpublished report).

3 It is important to bear in mind the limitations of this approach to grouping soil types. Physical properties of the soil (such as moisture holding capacity), as well as previous management history, are important sources of variation in the response to fertilizer application among soil types of similar chemical properties. The more appropriate way of accounting for the effects of soil type would be to quantify soil fertility levels by measuring base levels of N and P every season in all sites. Because adequate data were lacking on base levels of nutrients in the soil for each site and year, the present study could use only qualitative groupings to control for the effects of soil type.

4 The coefficient of this soil type is obtained from the rule $\Sigma_i\ d_i = 1$, where d_i is the deviation of the mean effect of category i from the overall mean. This rule is employed to drop one argument to avoid singularity of the $x'x$ matrix when all dummies are estimated together with a constant term.

5 Tea and coffee growers were provided with fertilizer through the KTDA and other cooperatives at less than 60% of the market price during the 1993/94 season (Government of Kenya, unpublished report).

6 The specification of the logit and tobit models and their applications to the analysis of technology adoption in general are discussed in more detail in Chapters 4 and 7.

9 Determinants of the Incidence and Severity of *Striga* Infestation in Maize in Kenya

RASHID M. HASSAN AND JOEL K. RANSOM

Introduction

Striga **species (witchweeds) are parasitic weeds causing increasing losses in cereal crops, including maize, throughout sub-Saharan Africa. Results of the survey of maize farmers, as well as other data, were used to map the spread and severity of *Striga* infestation in maize in Kenya. A model was developed to measure and test hypotheses about the causal relationship between *Striga* infestation and a set of explanatory variables. Results from recent field experiments conducted in western Kenya were compared with these relationships. The analysis confirmed that the incidence of *Striga* in maize is increasing at the margins of Kenya's largest maize-producing agroecology, the moist transitional zone. High priority must be given to research and extension focusing on methods to prevent *Striga* from spreading into this zone. The regression analysis supported research findings on the effects of agroecological zone, soil type, population pressure and intensification, cropping intensity, method of weeding, and soil fertility level on the development of *Striga* in maize. On the other hand, no conclusive evidence emerged that local versus improved maize, time of planting, and cropping pattern either encourage or discourage *Striga*. Further research is needed on maize varieties that tolerate *Striga* infestation and on assessing whether monocropping or mixed cropping (and separate or mixed row intercropping), are better for managing *Striga*. The effects of different intercrops on *Striga* infestation should continue to be studied as well. Finally, it is very important to conduct more on-farm research to tap farmers' knowledge on the control of *Striga* and to ensure that farmers will be able to adopt any innovations that are developed.**

Striga species, commonly known as witchweeds, are root parasites and a serious constraint to cereal production in sub-Saharan Africa. *Striga* is estimated to infest 21 million hectares and could spread to an additional 23 million hectares (Sauerborn 1991). Within the genus *Striga*, two species are the most prevalent and damaging to agricultural systems: *S. hermonthica* and *S. asiatica. Striga hermonthica,* the most vigorous and pernicious, is found throughout semiarid areas of northern tropical Africa (Parker and Riches 1993). The weed is indigenous to East Africa, where it attacks all of the major tropical cereal crops and sugarcane. Although *S. hermonthica* can be found under wild conditions (Doggett 1965), it develops most often on cropped cereals. *Striga asiatica*, which is found throughout Africa, parts of Asia, and the USA, is generally less aggressive and damaging than *S. hermonthica.* Within Kenya, *S. asiatica* grows along the coast (Terry and Michieka 1987) and in isolated areas of western Kenya. Unlike *S. hermonthica, S. asiatica* most often parasitizes wild vegetation rather than cropped species.

In Kenya, maize is one of the most important cereal crops threatened by increasing populations of *Striga.* Maize accounts for 80% of the national production of cereals and is a significant source of carbohydrates for Kenyans, who consume more than 100 kg of maize per person each year (CIMMYT 1994). Although the importance of maize relative to other cereals has not been documented in the zones where *Striga* is a problem, maize generally is the dominant subsistence crop in Kenya. This dominance tends to exacerbate the *Striga* problem, as commercially available varieties of maize are more susceptible to *Striga* than commercially available varieties of sorghum, the alternative subsistence cereal crop (Ransom and Odhiambo 1992).

Striga clearly is adapted to a wide range of environmental conditions in tropical Africa, and methods of preventing the weed from spreading further are needed urgently. Low soil temperature has been identified as a constraint to *Striga* in environments with sufficient moisture to allow for cereal production (Patterson 1987). *Striga* is a pest of agricultural intensification, associated with increased cropping intensity and declining soil fertility (Parker 1991). The application of nitrogen fertilizer has been shown to reduce the level of *Striga* infestation (Pieterse and Verkleij 1991), but once *Striga* populations have built up to high levels, the returns to nitrogen fertilizer alone are insufficient to encourage farmers to use fertilizer to control the weed (Ransom and Odhiambo 1994).

Research on methods for controlling *Striga* has been hampered by a lack of precise information on factors affecting the extent and severity of the weed problem. Although *Striga* has been identified as one of the major constraints to maize production in narrowly focused on-farm surveys done in a few areas of western Kenya, little local information is available on factors affecting populations of the weed. Except for two very recent surveys (Frost 1994; Hassan *et al.* 1995), no information has been published on the extent of *Striga* in Kenya or the factors conducive to its buildup and spread.

This chapter describes results of a study done to supply more information to researchers on which biophysical and socioeconomic factors influence the spread and severity of *Striga* infestation. The study also sought to investigate the causal relationships between these factors and incidence of the weed. A third objective of the study was to identify locations in Kenya where additional agronomic research is likely to develop improved *Striga* control options for farmers.

Methods

Several sources of information were used in the analysis, including the data defining the maize-specific agroclimatic zones and gathered in the national survey of maize farmers (Chapters 2, 3, and 4), as well as data from research on *Striga* in Kenya. The extent and severity of *Striga* infestation in maize were mapped using spatial analysis. A formal model was developed to measure and test hypotheses about the causal relationship between *Striga* infestation and a set of explanatory variables. Results from recent field experiments conducted in western Kenya were compared with the modeling results.

The KARI maize program initiated field experiments on *Striga* in 1989. This research, conducted in collaboration with CIMMYT, was largely funded by the Canadian International Development Agency (CIDA) and the CIMMYT East Africa Cereals Project. Trials were conducted on the experiment station of the National Sugar Research Center, Kibos, and at KARI subcenters in Alupe and Kibos. These three sites represent diverse soil types within the zone of serious *Striga* infestation in western Kenya. Some on-farm experiments were conducted as well.

Results and Discussion

The spread of Striga *in maize-producing areas*

The geographical distribution of *Striga* is depicted in Geographical Data Map 3, based on survey data on the percentage of maize area infested with the weed. *Striga* is most common in the relatively warmer (lower elevation) loca-

Table 9.1. Distribution of *Striga* infestation on maize in Kenya, by agroclimatic zone.

Agroclimatic zone	Elevation (m)	Average total seasonal rainfall (mm)	Daily temperature (°C) Min.	Daily temperature (°C) Max.	Area sown to maize (000 ha)	Percentage of maize area infested with *Striga*	Average percentage damage caused
Lowland tropics	<800	<1000	20.0	29.4	33	0	0
Dry midaltitude	700–1300	<600	16.1	27.9	118	0	0
Moist midaltitude	1100–1500	>500	15.9	28.3	118	39	51
Dry transitional	1100–1800	<600	14.0	25.3	37	0	0
Moist transitional	1200–2000	>500	13.4	23.3	424	1	42
Highland tropics	>1600	>400	10.0	23.0	307	0	0
Total					1037	2	48

Source: Survey of maize farmers.

tions of western Kenya, mainly the moist midaltitude zone around Lake Victoria. On average, 39% of the 118,000 ha planted to maize in the moist midaltitude zone (which accounts for 12% of Kenya's maize area) is infested with *Striga* (Table 9.1). However, *Striga* is also moving into the moist transitional zone, Kenya's largest maize-producing region (accounting for 40% of national maize area), and this movement is an issue of great strategic significance for researchers and policy makers. The fact that the area of maize infested with *Striga* is small relative to the area that could potentially become infested indicates that preventive work to contain *Striga* is even more important than 'curative' efforts to control the weed in infested areas. An effective means of preventing the weed from spreading would not only save the 400,000 ha of maize under immediate threat in the moist transitional zone, but would keep *Striga* out of the midaltitude areas to the east. While *Striga* is not yet reported in the drier segments of the midaltitude and transitional zones east of the Rift Valley (Map 3), moisture levels in those zones are considered sufficient for *Striga* to develop there (Hassan *et al.* 1995). Fortunately, the dry and moist segments of the midaltitude and transitional zones are separated by a wide area of the highland tropics, where the climate is too cold and wet for *Striga.* This geographical barrier and the manner in which *Striga* was introduced in Kenya, rather than climatic factors *per se*, are believed to be responsible for the weed's localized infestation in western Kenya. However, if Striga is not contained, it will be only a matter of time before it creeps into other zones.

Farmers in the coastal areas of Kenya did not report that *Striga* was a problem, although *S. asiatica* was previously reported in these areas. The fact that *S. asiatica* was not identified as a serious constraint suggests that it is less aggressive and less likely to spread, at least within the traditional farming systems of this environment. Cereal production in the coastal areas is less intensive than in other zones, and cereal crops are grown on farms with large areas devoted to tree crops, which may also explain the more localized infestations of *S. asiatica* in coastal areas compared to *S. hermonthica* in western Kenya. Furthermore, the maize survey is likely to have missed regional patterns that are highly localized. A more intensive, *Striga*-specific survey found *S. asiatica* to be very severe in localized patches along the coast (Frost 1994).

Determinants of severity of Striga *infestation*

Several factors affecting the incidence and severity of *Striga* infestation have been reported in the literature, and Weber *et al.* (1993) classified these factors into primary and modifying determinants. The most important primary factors were climate, soil, cropping pattern, and cropping intensity; factors that modified the incidence of *Striga* included plant variety and cultural practices. *Striga* was found to be confined to particular climatic conditions, such as the subhumid and semiarid areas of the tropics (Weber *et al.* 1993), and high levels of infestation were associated with shallow soils, more intensive farming, and low soil fertility (Patterson 1987; Lagoke *et al.* 1991; Parker 1991; Vogt *et al.* 1991). Time of planting and weeding were important modifying determinants of *Striga* infestation (Lagoke *et al.* 1991). To examine the causal relationship between *Striga* and several of these determining factors, a formal model was developed and estimated; the model and results are described next.

The empirical model

The model was fitted to data collected during the survey of maize farmers (Chapter 3). The dynamics of *Striga* populations are usually monitored at different stages of infestation. Seed counts during soil infestation and *Striga* plant counts during emergence, the reproductive stage, and seed production are usually taken (Weber *et al.* 1993). The extent of infestation in each stage is a function of population levels in the preceding stage. Although such data were not obtained through the farmer survey, enumerators did ask farmers whether *Striga* was an important problem in their maize fields and asked them how much damage *Striga* did to the maize crop.[1]

Farmers' responses were used to define measures of the incidence and severity of *Striga* infestation in the survey area. Farmers' estimates of the percentage yield loss caused by *Striga* (y) is used as a proxy measure of severity.[2] Variable y is censored (or truncated), in the sense that it is observed in only a limited range. For instance, the severity or extent of the damage (i.e., value of y) is zero (or unobserved) for farms not infested with *Striga*. The censored regression model is specified as follows:

$$y = X\beta + \varepsilon \quad \text{if } y > 0 \text{ (i.e., \textit{Striga} is a problem);}$$
$$\text{otherwise, } y = 0 \text{ (i.e., \textit{Striga} is not a problem),} \qquad (9.1)$$

where y measures severity, X is a vector of known constants, β is a vector of unknown parameters, and ε is the random error term. Using the Ordinary Least Squares (OLS) method to estimate model 1 will generate biased and inconsistent estimates of β and σ_ε (variance), as the E[ε] # 0 in this case, because the dependent variable is truncated (Maddala 1983). The tobit model, however, is generally used to obtain consistent estimates of model parameters when the range of the dependent variable is limited,[3] and for this study we used the tobit procedure to estimate the parameters of equation (9.1). The maximum likelihood (ML) estimate is easy to compute in the tobit model, which estimates censored and truncated normal distributions (Maddala 1983).

Information on the following explanatory variables was also collected during the survey:

Physical climate. Variations in climate are captured in the maize-specific agroclimatic zones. Data in Table 9.1 and Geographical Data Map 3 show that *Striga* was reported only in the moist midaltitude zone and adjacent regions of the moist transitional zone (e.g., Kakamega, Nyamira, and South Nyanza Districts).

Soil. Soils in farmers' fields were not sampled during the survey, but enumerators collected general information on the dominant soil type at each survey site. This information was used to classify soils into mainly sandy and mainly clay groups, as a proxy for soil texture and depth.

Cropping system. Three variables were defined to examine the effect of cropping system on *Striga* infestation:

- *System* (a binary variable with the value of one if mixed cropping of maize is followed and zero if monocropping is practiced).
- *Type of intercrop* is also assumed to influence *Striga.* Intercrops are therefore classified into cereals (value of one) and noncereals (zero). Noncereals such as legumes were observed to suppress *Striga,* whereas cereals such as sorghum serve as co-hosts with maize (Singh 1990; Parker and Riches 1993).
- *Field arrangement for intercrops.* Farmers use several field arrangements when they interplant maize with other crops. Intercrops may be planted between maize rows (separate rows), between maize plants in the same row (mixed rows), or in the same hole with the maize seed. It has been reported that intercropping in mixed rows results in lower infestation than intercropping in separate rows (Weber *et al.* 1993). A binary variable with the value of one for between-row planting and value of zero otherwise is constructed to test for the effect of intercropping arrangements.

Intensity of land use. Cropping intensity was represented by two factors:

- *Rotation 1 (interyear).* This variable is the number of years that land is continuously cropped with maize. Crop rotations are assumed to determine the availability of suitable hosts for *Striga* (Weber *et al.* 1993). Including more fallow periods in the rotation or alternating maize with other crops is expected to reduce *Striga* infestation.
- *Rotation 2 (intrayear).* This variable takes a value of one if maize is double-cropped and zero if it is not (single-season maize).

Population pressure. This variable is measured as the number of people per square kilometer. It is hypothesized that population pressure affects the incidence of *Striga* in two ways: as a source of more labor for weeding (which would suppress *Striga*) and as a source of increased cropping intensity (which would favor the development of *Striga*).

Maize variety. This variable controls for the effect of variety (value of one if a local variety is used and zero if an improved variety or hybrid is used). Farmers claim that local cultivars tolerate *Striga* better than improved maize (Hassan *et al.* 1995).

Time of planting. This variable is measured as weeks from the beginning of the year.

Use of basal fertilizer (Yes, No).

Use of animal manure (Yes, No).

Time of planting, use of basal fertilizer, and use of animal manure are included to test hypotheses and verify results obtained elsewhere on the use of organic and inorganic fertilizers and delayed planting to suppress *Striga* (Lagoke *et al.* 1991; Ariga and Berner 1993; Weber *et al.* 1993).

Method (mechanical/chemical, or manual) and number of weedings.

Results of the empirical analysis

As noted earlier, the effects of this set of variables (X) on *Striga* infestation (y) are analyzed and measured using the tobit model. The analysis confirmed earlier findings about the effect of climate on *Striga* (Table 9.1), as the percent-

Table 9.2. Parameter estimates of the tobit regression model for percentage yield loss resulting from *Striga*.

Explanatory variable	Regression coefficient	*t*-ratio[a]
Moist midaltitude zone	3.23	1.76*
Population density (person km^{-2})	0.01	0.80
Soil type (mostly sandy)	0.90	0.61
Variety (Lake locals)	3.85	2.04**
Cropping system (intercropping)	3.67	1.98**
Intercropping arrangement (separate rows)	−0.61	-0.82
Type of intercrop (cereals)	0.59	0.34
Intensity of cultivation (double cropping)	2.79	1.84*
Years of continuous maize cultivation	0.04	1.00
Basal fertilizer (used)	−0.88	-0.45
Animal manure (used)	−1.34	-0.86
Manual weeding	−1.45	-0.78
Number of weedings	−1.33	-0.54
Sowing date (weeks from beginning of year)	−0.08	-1.79*
Constant	18.13	2.24**
Number of cases 262		
R^2 0.26		
F-ratio 2.4**		

[a] * and ** denote statistical significance at 10% and 5% levels, respectively.

age yield loss was estimated to be higher in the moist midaltitude zone than in the moist transitional zone (Table 9.2). Though the results for the population density variable were not statistically significant, higher population densities were found to increase the severity of *Striga* infestation. The cropping intensification resulting from increased population pressure (which favors growth of the weed) had a stronger effect on *Striga* than the increased availability of labor for weeding (which would have suppressed rather than encouraged *Striga*).

For the soil type variable, results of the analysis, though not statistically significant, were similar in direction to results of agronomic research in northern Ghana (Vogt *et al.* 1991): *Striga* appeared to cause more damage on sandy than on clay soils. However, other research has suggested that soil type does not have a direct effect on the biology of the parasite. Observations from trials in locations with varying soil types in western Kenya have established that *Striga* grows, develops, and increases in intensity on all major soil types, including the heavy vertisol at Homa Bay (Ransom and Odhiambo 1994). Even so, Ransom and Odhiambo have emphasized that sandy soils are more likely to lose soil nitrogen to leaching and have higher rates of organic matter oxidation, which may encourage *Striga* populations to build up, just as better soil nitrogen levels and organic matter content have been shown to arrest the buildup of *Striga*.

The tobit analysis did not bear out farmers' claims that local maize cultivars possess better tolerance to *Striga*. *Striga* was found to cause more damage to local maize than to improved maize, and the effect of variety was statistically very significant (Table 9.2). The survey data indicated that most maize farmers in the moist midaltitude zone, where *Striga* is a serious weed, were using local varieties (Chapter 7). Either farmers' estimates of yield losses are inaccurate, or varietal attributes other than tolerance to *Striga* attract farmers to the local cultivars. Research in western Kenya indicates that commercial varieties and hybrids available in the region vary considerably in the levels of *Striga* that they can tolerate, though most of these differences can be explained by differences in time to maturity of the commercial cultivars. Full-season cultivars tolerate *Striga* better than cultivars with a shorter field duration (Ransom and Odhiambo 1995). Research in progress suggests that the local cultivars which farmers regard as resistant to *Striga* support levels of infestation similar to those borne by commercial cultivars with the same time to maturity. Local maize materials may possess some tolerance which allows them to yield slightly more than commercial materials when attacked by *Striga* (Ransom and Odhiambo 1992).

Cropping system factors were found to be very important determinants of the severity of *Striga* infestation. Contrary to earlier findings (Parker and Riches 1993), results of this analysis showed that intercropping increased the reported severity of *Striga* infestation (Table 9.2), although planting maize and other crops in separate rows resulted in less damage than intercropping in mixed rows. As expected, yield losses were lower when maize was cropped with noncereal crops rather than other cereal crops. Research in western Kenya, which was limited to investigating the effect of cowpeas intercropped with maize, has indicated that intercropping generally reduced *Striga* (Odhiambo and Ransom 1991).[4] Between-row intercropping was more conducive to *Striga* than within-row intercropping, but even when *Striga* numbers were reduced, maize yield did not increase, possibly because of the heavy competitive effect of the cowpeas. Cowpea yields were minimal as a result of severe damage by insects.

More intensive farming (planting two maize crops every year; less fallow or more years of continuous maize cropping on the same piece of land) resulted in greater damage from *Striga* (Table 9.2). Fallowing affects *Striga* both directly (by the absence of a host crop for the weed) and indirectly (by reducing the extraction of nutrients from the soil). Some of these beneficial effects would probably not be evident for a number of years, however, and the intensity of cropping in western Kenya makes it unlikely that farmers would be able to fallow for longer than three to four seasons. Trials over several seasons in western Kenya have shown that fallowing did not consistently reduce *Striga* seed numbers in the soil compared with continuous maize cropping, even though the natural grasses in the fallow did not support *Striga* and allow it to reproduce (Odhiambo and Ransom 1995). Furthermore, fallowing produced no yield advantage for maize; maize grown after three years of fallow yielded no better than maize grown continuously on the same land season after season (Odhiambo and Ransom 1994). Rotating maize with fertilized cotton, however, reduced the number of *Striga* seeds in the soil compared with continuous maize cropping, fallowing, or a maize–cowpea rotation.

Although their effect was not statistically significant, inorganic fertilizers and animal manure were found to suppress *Striga* populations and reduce damage to the maize crop. This finding is at variance with data from research. Applications of inorganic and organic fertilizers over just one season have been found to produce little observed effect, but longer term studies have suggested that combinations of organic and inorganic sources of nitrogen are extremely important (Ransom and Odhiambo 1994; Odhiambo and Ransom 1995); they seem to have a strong additive effect in reducing *Striga*.

Survey farmers indicated that manual weeding controlled *Striga* better than mechanical or chemical means (this was not significant, however). Longer term research in western Kenya found that hand weeding reduced *Striga* populations in the following year only after three consecutive seasons (Ransom and Odhiambo 1994; Odhiambo and Ransom 1995). It is not surprising that farmers perceive little effect on *Striga* from hand weeding, given the relatively long period before *Striga* populations actually begin to decline. The tobit analysis found that, as one would expect, more weeding reduces the severity of *Striga* infestation (Table 9.2).

Finally, planting maize late appeared to reduce the severity of *Striga* infestation, although it is unclear why this should be so, and research results do not support this finding. Trials on research stations, in which planting was delayed a maximum of six weeks after the start of the rains, found that planting date did not significantly affect *Striga* populations (Ransom and Osoro 1991). Furthermore, delayed planting carries the risk of significant losses in maize yield as a result of drought stress late in the cropping season.

Implications for Future *Striga* Research

Perhaps the most important finding of this study is that the incidence of *Striga* in maize is increasing at the margins of Kenya's largest maize-producing agroclimatic zone, the moist transitional zone. High priority must be given to research and extension focusing on methods to prevent *Striga* from spreading into this zone. At this stage, it is more important to contain *Striga* within its current boundaries than to seek measures for controlling *Striga* in this zone.

Until research on *Striga* yields more conclusive results, specific preventive or curative recommendations cannot be formulated for farmers. However, the regression analysis described in this chapter confirmed the effects of some factors on the spread of *Striga* and provided some additional insight into which kinds of research might yield useful results. Despite the limitations of survey data and farmers' estimates, the regression analysis supported research findings on the effects of agroclimatic zone, soil type, population pressure and intensification, cropping intensity, method of weeding, and soil fertility level on the development of *Striga* in maize. Research results suggest that weeding and fertilization can be effective, but only in the medium to long term. However, intercropping and rotation of maize with noncereal crops have shown considerable potential for reducing the number of *Striga* seeds in the soil and subsequent levels of infestation.

On the other hand, our analyses produced no conclusive evidence that local versus improved maize, time of planting, and cropping pattern either encourage or discourage *Striga*. Further research is needed to investigate whether or not the local maize cultivars in the Lake Victoria area are more tolerant to *Striga* than improved materials. Researchers should also investigate whether mixed cropping or monocropping, and separate or mixed row intercropping, are better for managing *Striga*. Because intercropping is prevalent in *Striga*-infested areas, it is also important to study the effects of the various intercrops on *Striga* infestation. Intercropping maize with noncereal crops seems to have potential for reducing levels of *Striga* infestation. Finally, more on-farm research will be required to tap farmers' knowledge on the control of *Striga* and to ensure that farmers will be able to adopt any innovations that are developed.

References

Ariga, E., and Berner, D (1993) Response of *Striga hermonthica* seeds to different germination stimulants and stimulant concentrations. *Phytopathology* 83, 1401.

CIMMYT (International Maize and Wheat Improvement Center) (1994) *CIMMYT 1993/94 World Maize Facts and Trends. Maize Seed Industries Revisited: Emerging Roles of the Public and Private Sectors*. International Maize and Wheat Improvement Center (CIMMYT), Mexico City.

Doggett, H. (1965) *Striga hermonthica* in East Africa. *Journal of Agricultural Science* 65, 183–194.

Frost, H.M. 1994. Striga *Research and Survey in Kenya*. Final Report. KARI/ODA Crop Protection Project, Nairobi, Kenya. Mimeo.

Hassan, R., Ransom, J., and Ojiem, J. (1995) The spatial distribution and farmers' strategies to control *Striga* in maize: Survey results from Kenya. In: Jewell, D., Waddington, S., and Ransom, J. and Pixley, K. (eds) *Proceedings of the 4th Eastern and Southern Africa Regional Maize Conference*. International Maize and Wheat Improvement Center (CIMMYT), Harare, Zimbabwe, pp 250–254.

Lagoke, S.T., Adeosum, J., Ngawa, L., Iwuafor, E.N., and Nwasike, C. (1991) Effect of sowing date, nitrogen, and sorghum variety on *Striga hermonthica* in Nigeria. In: Ransom, J.K., Musselman, L.J., Worsham, A.D., and Parker, C. (eds) *Proceedings of the 5th International Symposium of Parasitic Weeds*. International Maize and Wheat Improvement Center (CIMMYT), Nairobi, Kenya, p. 534.

Maddala, G.S. (1983) *Limited Dependent and Qualitative Variables in Econometrics*. Cambridge University Press, New York.

Odhiambo, G.D., and Ransom, J.K. (1991) The effect of nitrogen and cowpea intercropping on development of *Striga* in maize fields. In: Kuria, J.N. (ed.) *Proceedings of the 13th Biennial Weed Science Conference*. Weed Science Society for Eastern Africa, Nairobi, Kenya, pp. 30–31.

Odhiambo, G.D., and Ransom, J.K. (1994) Preliminary evaluation of long-term effects of trap cropping and maize management on *Striga*. In: Pieterse, A.H., Verkleij,

J.A.C., and ter Borg, S.J. (eds) *Biology and Management of Orobanche, Proceedings of the 3rd International Workshop on Orobanche and Related* Striga *Research*. Royal Tropical Institute, Amsterdam, The Netherlands, pp. 505–512.

Odhiambo, G.D., and Ransom, J.K. (1995) Long-term strategies for *Striga* control. In: Jewell, D., Waddington, S., Ransom, J., and Pixley, K. (eds) *Proceedings of the 4th Eastern and Southern Africa Regional Maize Conference*. International Maize and Wheat Improvement Center (CIMMYT), Harare, Zimbabwe, pp. 98–113.

Parker, C. (1991) Protection of crops against parasitic weeds. *Crop Protection* 10, 6–22.

Parker, C., and Riches, C.R. (1993) *Parasitic Weeds of the World: Biology and Control*. CAB International, Wallingford, UK.

Patterson. D.T. (1987) Environmental factors affecting witchweed growth and development. In: Musselman, L.J. (ed.) *Parasitic Weeds in Agriculture. Vol. 1:* Striga. CRC Press, Boca Raton, Florida, pp. 27–47.

Pieterse, A.H., and Verkleij, J.A.C. (1991) Effects of soil conditions on *Striga* development: A review. In: Ransom, J.K., Musselman, L.J., Worsham, A.D., and Parker, C. (eds) *Proceedings of the 5th International Symposium of Parasitic Weeds*. International Maize and Wheat Improvement Center (CIMMYT), Nairobi, Kenya. pp. 329–339.

Ransom, J.K., and Odhiambo, G. (1992) Development of *Striga hermonthica* on maize and sorghum in western Kenya. In: Richardson, R.G. (ed.) *Proceedings of the First International Weed Control Congress, Vol. 2*. Weed Science Society of Victoria, Melbourne, Australia, pp. 430–433.

Ransom, J.K., and Odhiambo, G.D. (1994) Long-term effects of fertility and hand-weeding on *Striga* in maize. In: Pieterse, A.H., Verkleij, J.A.C., and ter Borg, S.J. (eds) *Biology and Management of Orobanche, Proceedings of the 3rd International Workshop on Orobanche and Related* Striga *Research*. Royal Tropical Institute, Amsterdam, The Netherlands, pp. 513–519.

Ransom, J.K., and Odhiambo, G.D. (1995) Effect of corn genotypes which vary in maturity length on *Striga hermonthica* parasitism. *Weed Technology* 9, 63–67.

Ransom, J.K., and Osoro, M.O. (1991) Effects of planting date and genotype on *Striga hermonthica* parasitism of maize and sorghum in Kenya. *Proceedings of the 12th International Plant Protection Congress*. Abstract 2 (1 August Session). International Plant Control Protection Congress, Brazil.

Sauerborn, J. (1991) The economic importance of the phytoparasites Orobanche and *Striga*. In: Ransom, J.K., Musselman, L.J., Worsham, A.D., and Parker, C. (eds) *Proceedings of the 5th International Symposium of Parasitic Weeds*. International Maize and Wheat Improvement Center (CIMMYT), Nairobi, Kenya, pp. 137–143.

Singh, R. 1990. Terminal report on sorghum and pearl millet agronomy. NCRE /IRA/IITA, Maroua, Cameroon, Mimeo.

Terry, P.J., and Michieka, R.W. (1987) *Common Weeds of East Africa*. FAO, Rome, Italy.

Vogt, W., Sauerborn, J., and Honisch, M. (1991) *Striga hermonthica* distribution and infestation in Ghana and Togo on grain crops. In: Ransom, J.K., Musselman, L.J., Worsham, A.D., and Parker. C. (eds) *Proceedings of the 5th International Symposium of Parasitic Weeds*. International Maize and Wheat Improvement Center (CIMMYT), Nairobi, Kenya, pp. 373–377.

Weber, G., Elemo, K., Awad, A., Lagoke, S.T., and Oikeh, S. (1993) Striga hermonthica *(Del.) Benth. in Cropping Systems of Northern Guinea Savanna*. Resource and Crop Management Research Monograph. International Institute of Tropical Agriculture (IITA), Resources and Crop Management Division, Ibadan, Nigeria.

Endnotes

1 Farmers' assessments were carefully verified through field observation by the enumerators, who were mainly maize researchers (breeders, agronomists, entomologists, and socioeconomists).

2 It is important to note that farmers' assessment does not represent the most accurate method of estimating yield losses related to *Striga*, mainly because several other factors influence yield, and their effects are difficult to isolate if not controlled for.

3 This model is often estimated using two-stage methods when the sample-separation criterion is of the probit type, e.g., the probit two-stage (Maddala 1983).

10 Availability and Effectiveness of Agricultural Extension Services for Maize Farmers in Kenya

Rashid M. Hassan, Danial D. Karanja, and H.H.A. Mulamula

A regression model was used to test hypotheses about possible determinants of the quantity (incidence) and quality (effectiveness) of extension services for maize farmers in Kenya. A particular effort was made to assess whether the Training and Visit (T&V) system of extension has been more effective than previous extension systems. Results of the analysis suggest that the T&V system was effective in reducing the bias of previous extension methods in disseminating information to more marginal production zones, small-scale and uneducated farmers, and remote areas. However, bias against female farmers persisted. Most farmers acquired information about new maize practices from extension sources, followed by learning from other farmers (possibly because of the increased use of contact farmers and farmer groups under the T&V system). Input supply systems contributed only minimally to the dissemination of new maize technology. The major constraint to adoption of improved maize varieties and pest control methods is lack of information, particularly in the lowland tropics. Efficient extension services and more efficient input supply mechanisms are required for wider diffusion and adoption of these technologies. The major barrier to increased adoption of fertilizer is not lack of information but high cost. Finally, in delivering research recommendations to farmers, more emphasis needs to be placed on marginal environments and female-headed households.

Agricultural extension is a common institutional instrument for promoting the adoption and diffusion of improved technologies developed through agricultural research. The original task of extension was simply to deliver research recommendations to farmers. These extension 'messages' were usually supported by credit arrangements giving farmers easy access to new inputs, often at subsidized prices. Two basic problems with this approach have been widely discussed in the extension literature. First, research recommendations were often not suited to farmers' needs; second, agricultural extension and credit services were directed mainly to farmers who had more resources and to farmers in environments more favorable for agriculture. The reasons often cited for these problems are that most agricultural research was removed from farmers' needs because it was conducted on experiment stations in isolation from farmers, farmers who were better off usually had collateral for repaying loans, and favorable environments had better infrastructure (Moris 1983; Feder and Slade 1986; Collinson 1987; Pirkhaeuser *et al.* 1991). The relevance of agricultural research to farmers' problems and circumstances has improved through adaptive research (the farming systems and participatory research approaches described, for example, in Collinson 1987; Farrington and Martin 1988; Chambers *et al.* 1989; Tripp 1991).

To enhance the effectiveness of extension and improve the productivity of agriculture, in recent years many countries have adopted a system called 'Training and Visit' (T&V) extension (Benor *et al.* 1984; World Bank 1985; Feder and Slade 1986; Dejene 1989; and Hussain *et al.* 1994). This chapter examines how the extension of technologies to maize farmers has changed in Kenya. The objective of this work is to inform maize researchers and extension policy makers of the strengths and deficiencies of current programs and services supporting technology dissemination among Kenyan maize farmers.

Data and Methods

In the survey of maize farmers, described in Chapter 3, enumerators asked farmers whether extension advice was provided to them at all, the year in which they had their first contact with the extension service, and their sources of advice on maize production. They were also asked whether they had adopted practices recommended by extension agents and, if not, why. Data were also gathered on the state of road and marketing infrastructure near survey farmers' villages as well as the availability of extension support services. Extension services associated with four major maize production practices were investigated: variety, fertilizer, pest control, and plant arrangement (spacing) recommendations.

These data were used to test hypotheses about the influence of several factors on the availability and effectiveness of extension services for maize farmers: agroclimate; farm size; farmers' characteristics, such as age, gender, and education; and infrastructure. Qualitative response models were used to measure and test hypotheses about the effects of each of those factors.

In analyzing the quantity and quality of extension services, it is important to take into account the effect of the T&V system, launched in Kenya in 1982. The T&V approach was designed to improve the effectiveness of extension services by overcoming basic deficiencies of conventional extension systems, such as poor training of extension agents, irregular work schedules, irregular visits to farmers, and poor links between research, extension, and farmers (Benor *et al.* 1984; World Bank 1993). The T&V system:

- Made use of frontline extension workers, contact farmers, farmer groups, subject matter specialists, and research scientists.

- Provided training every two weeks to frontline extension workers by subject matter specialists and every month to subject matter specialists by research scientists.
- Fixed work programs and regular visits (usually fortnightly) by extension workers to contact farmers or farmer groups.

The survey data offer a unique opportunity to assess T&V strategies in Kenya, where the new extension system was considered so successful that it was used to make the case for T&V elsewhere in Africa (World Bank 1990, 1993).

Arrival of Information and Extension Advice

In studying any extension system, it is important to identify which factors determine whether information on improved production methods reaches farmers. Table 10.1 presents data on how the diffusion of extension information varies across the agroclimatic zones for maize production in Kenya. Although more than 50% of the maize farmers in Kenya received information about improved production practices, significant variation exists from one zone to another. The extension service reached more farmers in high-potential zones (the moist transitional and highland tropical zones had the highest

Table 10.1. Exposure of Kenyan maize farmers to extension advice, by time of arrival, type of message and agroclimatic zone (% farmers).

	Agroclimatic zone						
Type of message	Lowland tropics	Dry midaltitude	Moist midaltitude	Dry trans'l	Moist trans'l	Highland topics	Total
Advice on variety							
Never received	58	52	41	44	21	39	37
Received before T&V	3	8	10	9	17	13	13
Received after T&V	11	18	17	10	26	21	20
Received, but don't remember date	28	22	31	38	35	27	30
Advice on fertilizer							
Never received	64	53	44	45	26	43	41
Received before T&V	2	6	9	8	17	12	11
Received after T&V	9	15	16	10	22	16	17
Received, but don't remember date	25	26	31	38	35	29	31
Advice on pest control							
Never received	63	52	45	46	27	44	42
Received before T&V	2	6	9	9	17	10	11
Received after T&V	11	17	14	9	21	15	16
Received, but don't remember date	24	25	32	36	35	30	31
Advice on plant arrangement							
Never received	63	52	45	46	26	41	41
Received before T&V	3	7	10	9	18	12	12
Received after T&V	9	18	15	10	23	17	17
Received, but don't remember date	25	24	30	35	33	30	30

Note: 'T&V' = the Training and Visit system of extension. 'Before T&V' = before 1982, 'after T&V' = from 1982 onwards.

rates of extension contact) than in relatively more marginal environments, such as the lowland tropics and semiarid areas. This was true for each type of extension message examined here – messages related to variety, fertilizer, pest control, and plant arrangement (spacing). These results are consistent with findings of other studies, which suggest that agricultural services in Kenya are relatively biased towards high-potential regions such as the Rift Valley and the western provinces (Chapters 7 and 8; Gerhart 1975). The high concentration of agricultural services in these areas has played a key role in the high levels of adoption of new maize technologies and the emergence of a modern maize sector.

More farmers received extension advice on improved varieties than on fertilizer use, pest control, and plant arrangement, reflecting the effectiveness and advantages of an organized public network for multiplying and disseminating seed, especially at early stages of agricultural development. In all zones and for all four kinds of practices, more farmers received advice after the T&V system was introduced (Figs 10.1 and 10.2).[1] However, it is important to note that about one-third of the farmers could not remember when they first became aware of the new practices (Table 10.1).

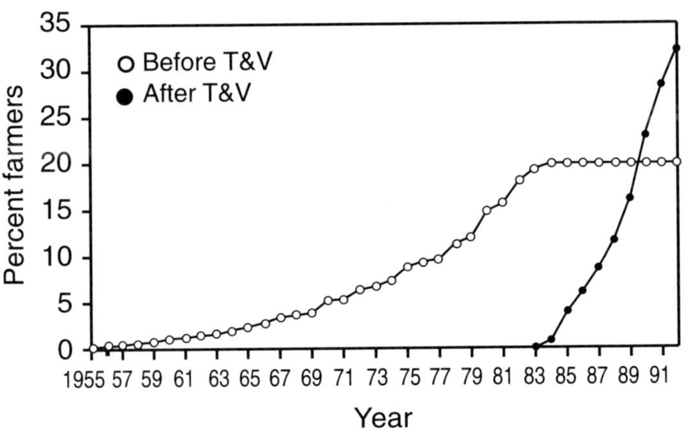

Fig. 10.1. Arrival of advice on variety before and after the T&V system, Kenya.

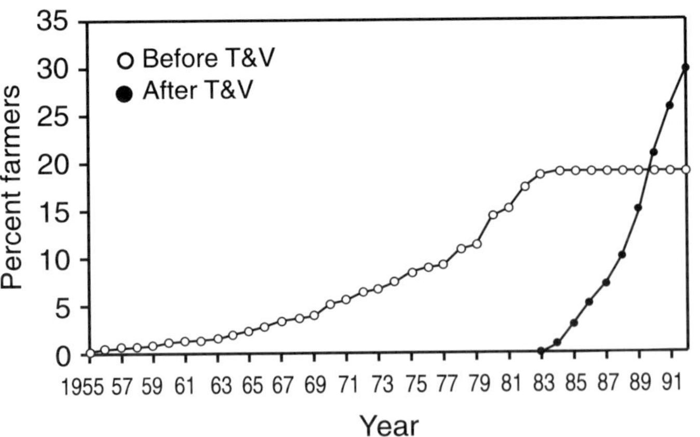

Fig. 10.2. Arrival of advice on fertilizer before and after the T&V system, Kenya.

Table 10.2. Exposure of Kenyan maize farmers to extension advice, by time of arrival, technology message, farm size, and gender of farmers (% famers).

Message	Farm size			Gender of farmer	
	Small (< 2 ha)	Medium (2–8 ha)	Large > 8 ha)	Male	Female
Advice on variety					
Never received	40	37	25	35	42
Received before T&V	9	13	26	15	9
Received after T&V	19	22	22	23	16
Received, but don't remember date	31	27	28	27	33
Advice on fertilizer					
Never received	43	43	28	38	46
Received before T&V	8	12	22	13	8
Received after T&V	17	18	15	19	13
Received, but don't remember date	32	27	35	30	32
Advice on pest control					
Never received	45	41	30	40	46
Received before T&V	8	10	22	12	8
Received after T&V	15	19	13	19	12
Received, but don't remember date	31	29	36	29	34
Advice on plant arrangement					
Never received	44	41	26	39	44
Received before T&V	8	14	22	14	8
Received after T&V	18	18	17	20	14
Received, but don't remember date	30	27	34	27	33

Note: 'T&V' = the Training and Visit system of extension. 'Before T&V' = before 1982, 'after T&V' = from 1982 onwards.

The likelihood that a farmer would receive information about improved practices increased as farm size increased (Table 10.2). It also increased for male farmers compared with female farmers. On the other hand, the percentage of large-scale farmers who received extension advice was lower after the T&V system was instituted, for all types of extension messages (Table 10.2), whereas the percentage of small- and medium-scale farmers receiving advice increased. This result suggests that the T&V system has contributed to reversing the bias against small-scale farmers. The bias against female farmers, however, has remained unchanged. After the T&V system was initiated, the percentage of male farmers receiving extension advice increased more than the percentage of female farmers (Table 10.2).

A formal model was developed to test hypotheses about the effects of these factors on farmers' awareness of and access to information on new maize production practices. Logistic regression analysis was used to investigate these relationships. The logit model, which measures the effect of determining factors on the probability that an event will occur (i.e., whether a farmer received extension advice), is expressed algebraically as:

$$P = 1 / (1 + \exp(Z)) \quad (10.1)$$
$$Z = X' \beta, \quad (10.2)$$

where P is the probability that an event (e.g., that a farmer will receive extension advice) will occur, X is the set of regressors, and β is the vector of parameters. When parameter β_i (coefficient on regressor x_i) is equal to zero, then Z is also zero, and hence $\exp(Z)$, which measures the odds of occurrence (defined below as O_i) will be equal to one. A unit value of O_i means that factor i has no effect on the odds that the event will occur.

$$O_i = \text{probab.(event occur)/probab.(event does not occur)}$$
$$= \exp(Z) \quad (10.3)$$

O_i will be greater than one for positive values of β_i, indicating that the odds of occurrence increase with x_i. The reverse is true for negative values of β ($\beta_i < 0$).[2]

Results of the logit analysis for the dichotomous dependent variable ADVICE (received = 1, not received = 0) are presented in Table 10.3. To test for the effect of changing to the T&V system, the sample is separated into two groups (the percentage of farmers contacted by extension before and after T&V), and the logit model is applied separately to each group.[3]

In addition to different farmer characteristics, agroclimatic zone, and infrastructure, the model includes tea and/or coffee planting as a separate variable. Compared to the Ministry of Agriculture's general extension services, extension services for tea and coffee production are the oldest and best developed in Kenya, owing to the highly commercial nature of tea and coffee production and the fact that these services must reach relatively small numbers of specialized farmers (World Bank 1993). It is hypothesized that maize farmers have a greater chance of obtaining extension advice if they also produce tea or coffee.

Results of the logit analysis indicate that the odds of getting extension advice before T&V were lower (negative β_i) for farmers in the more marginal (or smaller) zones, such as the lowland tropical, dry midaltitude (semiarid), and moist transitional zones, than in the high potential zones. Although the bias against the lowland tropical and dry midaltitude zones was maintained after T&V (negative β_i), the magnitude of the regression coefficient (β_i) was smaller, indicating that the odds of receiving extension advice had nevertheless improved. In other words, a lower negative value of β_i means that the odds factor, $O_i = \exp(Z)$, is larger.

The bias against the moist midaltitude zone, on the other hand, was clearly reversed after T&V (β_i became positive). The emphasis on the largest zone, the moist transitional zone, increased after T&V (larger positive values of β_i). While this is consistent with the T&V strategy of focusing on high-potential zones (World Bank 1993), coefficients switched sign (from positive to negative) in the highland tropics after T&V was instituted, indicating a reduced emphasis in another high-potential zone, except in the case of extension messages about pest control.

As with the effect for the moist transitional zone, the effects of farm size,

Table 10.3. Results of logit analysis of the determinants of access to extension advice among maize farmers in Kenya.

Regressors	Advice on variety		Advice on fertilizer		Advice on pest control	
	Before T&V	After T&V	Before T&V	After T&V	Before T&V	After T&V
Agroclimatic zone						
Lowland tropics[a]	−0.51	−0.18	−0.68	−0.32	−0.76	−2.13
Dry midaltitude	−0.510*	−0.488**	−0.594*	−0.44**	−0.62**	0.258
Moist midaltitude	−0.108	0.059	−0.022	0.111	−0.063	−0.54**
Dry transitional	0.441	−0.214	0.535	−0.040	0.835**	−0.019
Moist transitional	0.538**	0.89***	0.66***	0.88***	0.76***	1.1***
Highland tropics	0.149	−0.065	0.104	−0.188	−0.153	0.253
Farm size (large)	0.496***	0.289**	0.44***	0.267**	0.44***	0.299**
Gender (male)	0.075	0.162*	0.120	0.138	−0.021	0.25***
Age (years)	0.29***	−0.001	0.03***	0.005	0.03***	−0.005
Education (none)	−0.38***	−0.125	−0.45***	−0.193*	−0.45***	−0.082
Extension agent at nearest market	0.687***	0.53***	0.86***	0.47***	1.01***	0.317*
Distance to nearest market/center (km)	−0.001	0.001*	−0.001	0.001**	−0.001	0.001*
Plant tea or coffee	0.24	0.44***	0.26	0.28*	0.29*	0.41**
Constant	−3.1	−0.642	−3.7**	1.27**	−3.8**	−0.75*
Number of valid cases	667	678	682	721	681	557
−2 log likelihood	546	663***	473	640***	481	603**
Model χ^2	81***	66***	81***	55***	84***	67***

Note: figures in the table denote estimates of the regression parameter β_i. When $\beta_i < 0$, the odds factor O_i=exp(Z) < 1. The reverse holds for $\beta_i > 0$.

[a] To avoid singularity of the $x'x$ matrix, one dummy is dropped using the rule $_id_i = 1$, where d_i is the deviation of the mean of category i from the overall mean, and $i = 1, 2, \ldots N-1$, where N is the total number of dummies. The coefficient for the lowland tropics was accordingly derived to be the negative sum of the coefficients of the included zone dummies.

education, and proximity to extension services were all highly statistically significant. While the odds of getting extension advice continued to be higher (positive β_i) for large-scale farmers, the bias was reduced after T&V (smaller magnitudes). A similar comparison between male and female farmers suggests that T&V increased the odds of receiving advice for male compared with female farmers (except for pest control).

The probability of younger farmers receiving advice was lower before T&V than afterwards (either through switching sign or smaller magnitude of β_i). The same can be said about farmers with no education, as their chances of getting advice improved after T&V (through smaller values of β_i). Similarly, the bias towards farmers in areas where an extension agent was based nearby was also reduced after T&V (smaller positive values of β_i). Before T&V, the odds of getting advice decreased with distance from the nearest market center (Table 10.3). This trend was reversed after T&V, as β_i changed sign, indicating that more farmers in more remote areas had been reached. The logit results also supported the hypothesis of better access to extension among tea and coffee farmers across all technology messages. The effect of tea/coffee planting on the probability of extension contact was statistically significant.

In general, the results of Table 10.3 suggest that biases against smaller and relatively more marginal zones, small-scale farmers, younger and uneducated farmers, and farmers in remote areas were reversed by the T&V system, although the bias against women appeared to increase.

Farmers' Sources of Information about New Maize Technologies

The majority (about two-thirds) of surveyed maize farmers in Kenya who received advice on new methods acquired the information through extension, including visits of extension officers, field days, demonstration trials, and radio programs (Table 10.4). However, farmers learn about new technologies from various sources apart from the extension system. For instance, firms marketing agricultural inputs, such as seed, fertilizer, and pesticide, diffuse information about new inputs and farming practices to promote sales. Farmers also gather information about new practices by observing each other. Other farmers are the second most important source of information about new practices, particularly information on varieties (34%). The vast majority of farmers who acquired information from other farmers could not remember the year when they first became aware of the improved methods (96% in the case of varieties and 80–82% for the other technologies).

On the other hand, the proportion of farmers receiving advice through the extension system after the initiation of T&V was at least equal to the proportion receiving information before T&V (though it was slightly higher for varieties). It is likely that the number of responses in the 'other farmers' category reflects increased reliance on contact farmers and groups under the T&V approach. If this is the case, the impact of T&V may be underestimated. It is also clear from Table 10.4 that the contribution of input supply systems to disseminating maize technology has been relatively very low.

Factors Affecting the Efficacy of Advice on New Technologies

How effective is the advice that farmers receive about new technologies? Does this advice induce farmers to adopt recommended practices? Although data were lacking to assess the effectiveness of extension by measuring changes in productivity (Feder and Slade 1986; World Bank 1993; Hussein *et al.* 1994), the extent to which advice results in adoption served as a proxy measure of the efficacy of extension. The logit model was used to measure and test hypotheses about the relationship between the efficacy of advice and its determinants. Adoption or nonadoption of the recommended practices was used to index the binary dependent variable defining efficacy.

In addition to farmer characteristics, agroclimate, and economic infrastructure, the effectiveness of the T&V system on the quality of advice was

Table 10.4. Farmers' source of information on maize production practices, by type of message before and after the Training and Visit (T&V) extension system (% farmers).

Message	Extension[a]	Input supplier[b]	Other[c]
Improved variety			
Before T&V	91	8	1
After T&V	95	4	1
Don't remember	2	2	96
Total	62	4	34
Fertilizer use			
Before T&V	91	5	4
After T&V	91	3	6
Don't remember	8	12	80
Total	70	6	24
Pest control			
Before T&V	91	8	1
After T&V	94	4	2
Don't remember	18	2	80
Total	71	5	24
Plant arrangement			
Before T&V	93	3	4
After T&V	92	1	7
Don't remember	8	10	82
Total	71	4	25

Note: 'T&V' = the Training and Visit system of extension. 'Before T&V' = before 1982, 'after 'T&V' = from 1982 onwards.

[a] Extension agent, contact farmer, radio broadcast, field day, or demonstration trial.

[b] Trader/dealer selling agricultural inputs (e.g., seed, fertilizer, pesticides.

[c] Other farmers, chief, formal training.

compared with other systems and sources of information. The three logit models presented in Table 10.5 performed very well statistically. In general, low probability of adoption is associated with relatively marginal zones (the lowland tropical, dry midaltitude, and moist midaltitude zones) and with farmers who received no advice, had smaller farms, were older, and were uneducated. The odds of adoption also decreased for female farmers and farmers who were located far from extension services.

The effect of agroclimatic suitability was highly significant in the high-potential zones (the moist transitional zone and highland tropics). The source of advice and farm size effects were also very important, with a high statistical significance (Table 10.5). Results of the logit model showed that the odds

Table 10.5. Results of logit analysis of factors influencing adoption of the recommended maize production practice.

Regressors	Improved variety	Fertilizer use	Pest control
Agroclimatic zone			
Lowland tropics[a]	−1.74	−2.87	−2.57
Dry midaltitude	−0.05	−0.44*	0.31*
Moist midaltitude	−0.025*	−0.086	−0.52**
Dry transitional	0.16	0.74**	0.89***
Moist transitional	0.80***	1.19***	0.91***
Highland tropics	1.40***	1.29***	0.98***
Source of advice			
Extension, before T&V	0.58***	0.66***	0.53***
Extension, after T&V	0.63***	0.49***	0.87***
Other sources[a]	0.10	0.12	−0.07
Never received advice	−1.31**	−1.27***	−1.33***
Farm size (large)			
Gender (male)	0.52***	0.75***	0.75***
Age (years)	0.04	0.013	0.05
Education (none)	−0.005	−0.005	−0.006
Extension agent at nearest market	−0.122*	−0.025	−0.013
Distance to nearest market/center (km)	0.04	0.04	0.08
Plant tea or coffee	−0.001	−0.001**	−0.001***
Plant tea	−0.28**	0.32**	−0.263**
Constant	0.195	−0.149	−0.31
Number of valid cases	925	921	911
−2 log likelihood	971**	879	889
Model χ^2	310**	324***	287***

Note: figures in the table denote estimates of the regression parameter β_i . When $\beta_i < 0$, the odds factor $O_i = \exp(Z) < 1$. The reverse holds for $\beta_i > 0$. 'T&V' = the Training and Visit system of extension. 'Before T&V' = before 1982; 'after T&V' = 1982 onwards.

[a] To avoid singularity of the $x\,x$ matrix, one dummy is dropped using the rule $\Sigma_i d_i = 1$, where d_i is the deviation of the mean of category i from the overall mean, and $i = 1, 2, \ldots N-1$, where N is the total number of dummies. The coefficients for the lowland tropics and the other 'other source of advice' group were accordingly derived to be the negative sum of the coefficients of the included zone and source of advice dummies.

of adoption increased with the availability of information about the new practices (positive values of the regression coefficient for all sources). However, the magnitudes of the estimates for these parameters indicate that adoption was more likely to result from T&V extension, followed by extension services in general, than from the transfer of information from farmer to farmer.

Although planting tea or coffee did not promote adoption of recommendations related to maize variety or pest control, it did increase the odds that farmers would adopt fertilizer recommendations. This result confirms the findings of other studies, which have suggested that some of the fertilizer provided for tea or coffee farming (usually on credit or at subsidized prices) finds its way to the farm family's maize fields (Kimuyu *et al.* 1991; Murithi and Shiluli 1993). The lower probability that tea and coffee farmers would adopt recommended maize varieties or pest control methods may be attributed to the fact that these practices require expensive external inputs, which must be purchased through direct cash payments and cannot be financed indirectly through tea and coffee farming arrangements.

The efficacy of extension advice, however, is hard to separate from the quality and relevance of the advice (the content of the extension message). Unfortunately the survey of maize farmers did not collect details about the *content* of the information received from the various sources of advice for the four production practices. When improved varieties are promoted, the extension message may be fairly concise, whereas extension messages for crop management practices are usually more complicated. This is especially true for fertilizer recommendations, which are usually conditioned by nutrient source, dose, timing, and so forth. Results of the logit analysis suggested that the quality and content of messages received from extension agents were superior to information from other sources in promoting adoption. The results also suggested that large-scale farmers – who can pay for inputs, have received more education, and have better access to supply sources (distance to market) – were important forces for converting awareness (information) into adoption.

Reasons for Nonadoption

Maize farmers who chose not to adopt recommended practices were asked why they did so. Farmers' responses are reported by zone (Table 10.6), farm size, and gender (Table 10.7). It is clear from Table 10.6 that lack of awareness of new technology is the major constraint to adoption in the lowland tropics, followed by unavailability of inputs. Increased efficiency of extension and input supply systems would greatly increase the potential for adoption of improved maize technologies in the lowland tropics, particularly near the coast. In all other agroclimatic zones, poor availability of improved maize seed and pest control inputs, as well as the high cost of fertilizers, are farmers' major reasons for not adopting new technology. It is important to note that very few (less than 10%) of the farmers who were aware of plant spacing recommendations did not adopt them, and hence results for plant arrangement advice are not reported in Tables 10.6 and 10.7. In these zones, as in the lowland tropics, more efficient input supply systems are needed for improved seed and pest control technologies. The adoption of fertilizer would benefit from less expensive soil fertility management practices.

Similar results are obtained for the farm size and gender variables. Poor

Table 10.6. Farmers' reasons for not using the recommended maize production practice, by agroclimatic zone (% farmers).

Message	Agroclimatic zone						
	Lowland tropics	Dry midaltitude	Moist midaltitude	Dry trans'l	Moist trans'l	Highland topics	Total
Improved variety							
Not aware of practice	49	27	6	10	16	32	23
Input not available	30	39	52	44	47	42	43
Expensive	11	20	37	31	28	13	23
Not best practice/disagree	10	14	6	15	8	13	11
Fertilizer use							
Not aware of practice	49	17	7	13	11	15	17
Input not available	27	34	48	32	33	29	34
Expensive	21	43	43	42	52	42	42
Not best practice/disagree	3	6	3	3	4	15	7
Pest control							
Not aware of practice	51	20	7	12	13	16	18
Input not available	30	35	50	41	43	40	41
Expensive	14	40	35	35	39	28	33
Not best practice/disagree	6	5	8	2	5	15	9

Table 10.7. Farmers' reasons for not using the recommended maize production practice, by farm size and gender (% farmers).

Message	Farm size		Gender of farmer	
	Small (< 2 ha)	Large (> 2 ha)	Male	Female
Improved variety				
Not aware of practice	24	19	22	25
Input not available	42	51	49	36
Expensive	24	19	19	29
Not best practice/disagree	11	12	11	11
Fertilizer use				
Not aware of practice	17	15	17	17
Input not available	33	37	38	31
Expensive	43	36	40	45
Not best practice/disagree	7	12	8	7
Pest control				
Not aware of practice	18	16	17	20
Input not available	41	42	43	39
Expensive	32	35	31	34
Not best practice/disagree	9	7	10	7

availability and high cost are reported as the main constraints to adoption by both small-scale and large-scale farmers, as well as by male and female farmers (Table 10.7).

Conclusion: Implications for Research and Extension

Our analysis suggests that the T&V system has been effective in reducing the bias of previous extension methods in disseminating information to more marginal production zones, small-scale and uneducated farmers, and farmers in remote areas. However, bias against female farmers continued after the T&V system was introduced. Most farmers learned about new maize practices from extension, followed by learning from other farmers (which can possibly be attributed to the increased use of contact farmers and farmer groups under the T&V approach). On the other hand, input supply systems have made only a minimal contribution to the dissemination of new maize technology. There is some evidence that extension in general, and T&V in particular, is more effective than other sources of information in promoting adoption of new technology.

Whereas farmers say that lack of information is the major constraint to the adoption of improved maize varieties and pest control methods (particularly in the lowland tropics), the major barrier to increased adoption of fertilizers is their high price. Maize farmers engaged in tea or coffee production enjoyed better access to extension advice and had higher adoption rates, especially of fertilizers. Since tea and coffee farmers have better access to fertilizer through credit arrangements, these results imply that some of this fertilizer is used in their maize fields.

Results of this study also emphasize that the strategy for ensuring wider diffusion and adoption of improved technologies for maize production in Kenya is to develop a more efficient extension service (characterized by regular visits and increased use of contact farmers and farmer groups) and more efficient input supply mechanisms. More emphasis needs to be placed on marginal environments and female-headed households. Given that the use of fertilizer lags far behind the use of improved seed in Kenya, especially in high-potential zones (Chapter 8), and considering farmers' concern with the high price of fertilizers, it would appear that the search for relatively cheaper methods of soil fertility management remains a primary challenge for agricultural research and extension to realize significant gains in maize productivity in Kenya.

References

Benor, D., Harrison, J. and Baxter, M. (1984) *Agricultural Extension: The Training and Visiting System.* World Bank, Washington, DC.

Chambers, R., Pacey, A., Thrupp, L. (eds) (1989) *Farmer First: Farmer Innovation and Agricultural Research.* Intermediate Technology Publications, London, UK.

Collinson, M. (1987) Farming systems research: Procedures for technology development. *Experimental Agriculture* 23, 365–386.

Dejene, A. (1989) The Training and Visit, agricultural extension in rainfed agriculture: Lessons from Ethiopia. *World Development* 17, 1647–1659.

Farrington, J., and Martin A. (1988) *Farmer Participation in Agricultural Research: A Review of Concepts and Practices.* Agricultural Administration Unit Occasional Paper No. 9. Overseas Development Institute (ODI), London, UK.

Feder, G., and Slade, R. (1986) The impact of agricultural extension: The Training and Visit system in India. *World Bank Research Observer* 1(2), 139–161.

Gerhart, J. (1975) *The Diffusion of Hybrid Maize in Western Kenya.* Research Report for CIMMYT. International Maize and Wheat Improvement Center (CIMMYT), Mexico City.

Hussain, S., Byerlee, D., and Heisy, P. (1994) Impacts of the training and visit extension system on farmers' knowledge and adoption of technology: Evidence from Pakistan. *Agricultural Economics* 44(3), 428–442.

Kimuyu, P., Jama, M., and Muturi, W. (1991) Determinants of fertilizer use on smallholder coffee and maize in Muranga District, Kenya. *Eastern Africa Economic Review* 1, 1–11.

Moris, J. (1983) *Reforming Agricultural Extension and Research Services in Africa.* Agricultural Administration Network Discussion Paper No. 11. Overseas Development Institute (ODI), London, UK.

Murithi, F., and Shiluli, M. (1993) Effects of the liberalization of fertilizer markets on the distribution and use of fertilizer on food crop production: A study on Embu and Meru Districts of Kenya. In: Mwangi, W., Rohrbach, D., and Heisey, P. (eds) *Cereal Grain Policy Analysis in the National Agricultural Research Systems of Eastern and Southern Africa.* CIMMYT/SADC/ICRISAT, Addis Ababa, Ethiopia.

Norusis, M. (1991). *SPSS/PC+: Advanced Statistics 4.0 for the IBM PC/XT/AT and PS/2.* SPSS, Chicago, Illinois.

Pirkhaeuser, D., Evenson, R., and Feder, G. (1991) The economic impact of agricultural extension: A review. *Economic Development and Cultural Change* 3(39), 607–650.

Tripp, R. (1991) *Planned Change in Farming Systems:* Progress in On-Farm Research. Wiley, Chichester, UK.

World Bank (1985) *Agricultural Research and Extension: An Evaluation of the World Bank's Experience.* World Bank, Washington, DC.

World Bank (1990) *Kenya, Second National Agricultural Extension Project: Staff Appraisal Report.* World Bank, Washington, DC.

World Bank (1993) *Evaluation of the Performance of T & V Extension in Kenya.* World Bank, Washington, DC.

Endnotes

1 The T&V system was first introduced in Kenya in 1982 in only two districts. By 1985, the system had been extended to all districts covered by the MDBP survey (World Bank 1993). Farmers' responses to the survey questions on when and from which source they first received advice on the major maize production technologies were used to classify respondents who were contacted before 1982 as members of a pre-T&V group, versus a post-T&V group reached after 1982.

2 See Norusis (1991) for further details on the derivation of equation (10.3).

3 Results on plant arrangement advice were similar to those for varietal recommendations, and hence were not reported in Table 10.3 because of space considerations.

IV Conclusions

11 Current Challenges and Strategic Future Choices for Maize Research and Policy in Kenya: A Synthesis of the Maize Data Base Project Methods and Results

WILFRED MWANGI, JOHN LYNAM, AND RASHID M. HASSAN

This chapter recapitulates the major features of the spatial approach and GIS tools applied, the design and application of the spatially referenced database, and the main results of the MDBP. Strategic issues and future challenges related to sustained growth and transformation of maize production in Kenya, as well as their implications for research and policy design, are discussed. These include strategic issues related to the design and targeting of maize technology, technology support systems, and market liberalization – particularly adjustments in maize output markets, fertilizer markets, and maize seed systems. With regard to all of these issues, increasing the information flow is crucial to making the decisions that will lead to sustained growth in maize productivity. The MDBP offers a new base on which to evaluate the research plans and fine tune the market strategies that will lead to the next generation of productivity gains in Kenya's maize cropping systems.

Introduction

The successful strategy that transformed maize production in Kenya over the past three decades featured several critical elements: institutional innovation and investment in technology development; research; seed and fertilizer delivery systems; guaranteed prices and marketing; relatively developed road infrastructure; and large promotional extension campaigns. During that period, the status of maize in Kenya's agricultural economy deepened and maize became the most important food staple for all Kenyans. Maize area expanded and large yield gains were achieved, concurrent with smallholders' integration into the market system and the broader commercialization of maize production and distribution. However, the early success and growth fueled by this first generation of technical change, especially in smallholder maize production systems in the high potential areas, have faltered during the past decade. Further growth has been frustrated by several factors, including an inability to move beyond first-generation technical innovations; the pressure that population growth and farm subdivision placed on effective soil management; the expansion of maize production into more marginal environments; and the market uncertainty that has accompanied the liberalization process. As a result, yield gains have not kept pace with rapid growth in per capita demand for maize, and the country has become increasingly dependent on imports to close a recurrent deficit in domestic maize supplies, particularly during drought years. Maize researchers and policy makers have therefore been presented with new challenges that require them to rethink old strategies and design the next generation of technologies, institutions, and policies that will lead to sustained growth in maize productivity.

The MDBP was launched to gain insight into the factors responsible for early technological success and to establish a deeper knowledge and understanding of the nature and complexity of emerging challenges and constraints to sustained growth in maize production. The project took advantage of new tools for designing technology and for policy and research planning and evaluation, such as methods of spatial analysis and advanced computer capabilities in building and manipulating large digital databases. In the next section of this chapter, we recapitulate the major features of the spatial approach and GIS tools applied, the design and application of the spatially referenced database, and the main results of the MDBP. We then discuss the strategic issues and future challenges that are related to sustained growth and transformation of maize production in Kenya and their implications for research and policy design.

Study Approach and the Current State of Transformation in the Maize Sector

The spatial framework and design of the integrated digital database

The high-potential zones of Kenya were the target of the first-generation seed and fertilizer technology and the source of early productivity gains in maize. As maize production has expanded into relatively marginal environments over recent years, it has become increasingly important to understand the diversity in the biophysical and socioeconomic conditions shaping production in these newer maize adaptation areas. This understanding was crucial for proper definition of the biological potential of various agroclimates and for appropriate research design and effective targeting of technology, policy, and institutional innovations. Another important feature of transformation in Kenya's maize sector is the fact that a steadily increasing number of resource-poor, small-scale farmers in both high- and low-potential areas began adopting the improved high-input maize production methods designed and introduced to suit large-scale commercial farmers. It was therefore important to characterize the socioeconomic circumstances of smallholder maize systems and evaluate the suitability of currently promoted technology packages.

The spatial approach was adopted by the MDBP as the appropriate framework for properly defining production environments and capturing the spatial diversity in socioeconomic, infrastructural, and institutional conditions of maize farming across agroclimatic zones. The spatial approach also provided a suitable framework for designing georeferenced farmer surveys that would capture socioeconomic variability across farming households and for subsequently integrating survey information into the digital spatial database. Moreover, the spatial database provided a robust framework for systematically monitoring and assessing the adoption and impact of technology and a sound basis for *ex ante* research evaluation and priority setting. A longitudinal data set can easily be established through the monitoring system, allowing effective impact assessment. The current spatial frame and database can be linked to future data collection efforts designed to allow for a higher density of coverage, increased depth of measurement, as well as replication of the original inquiry into the state of technology diffusion.

Breeding for higher yielding ability has been the chief strategy pursued by maize technology development research and policy in Kenya. However, the selection and adaptation of germplasm at KARI was organized to serve broadly defined agroclimates separated on a crude basis of altitude and moisture availability. Prior to the MDBP, no map of well-defined, continuous boundaries providing a robust representation of the geographic distribution of maize agroclimates existed. Maize breeding research had no access to an adequate framework that could account for the wide range of biotic and abiotic stress factors across the country and thus permit the biological potential of available maize materials to be assessed with greater precision. The proper characterization of maize production environments and the development of a digital database and classification scheme for maize agroclimatic zones was the starting point and a principal product of the MDBP.

The initial zonation scheme developed by the MDBP used long-term data series on climatic attributes of seasonal relevance to maize production in Kenya to reproduce a refined digital version of the crude classification previously adopted by maize breeders. The major advantage of the new zonation system is that it is based on an interactive digital database that allows flexibility for future revision and redefinition of zonal boundaries as better or more information becomes available. The MDBP studies have exploited this dynamic feature of the digital maize database to address various technology design and research policy issues.

The maize adaptation zones delineated by the MDBP were separated on the basis of climatic suitability and experts' definitions of adaptability ranges for maize germplasm. However, biophysical suitability is not the only determinant of maize productivity. Several other farmer-specific, socioeconomic, institutional, and policy factors interact with the physical environment and influence the transfer of technology and its impact on productivity. For this reason, population-based sampling frames were combined with the spatial zonation scheme to capture socioeconomic variability within the defined climatic diversity. Population density and maize distribution intensity maps were overlaid on the initial spatial stratification of maize adaptation zones to determine sampling fractions within the climate–population clusters. A multistage, random cluster sampling was employed to select sites and farmers to survey. First, the national sampling frame employed for the most recent population census of 1989 was overlaid on the union of climate and population coverages to select survey sites using a random spatial search. Variable sampling fractions were determined proportional to intensity of maize cultivation (area under maize) and population density to provide proxy weights in the sample design for the two crucial impact measures. Systematic random sampling was then used to select farm households from sampled survey sites.

Applications of the spatial frame and database

The integrated digital database and spatial frame were applied to several aspects of the maize technology cycle and to research design, planning, and policy. These applications, presented in detail in previous chapters of this book, had several important results, which we summarize here.

Improved stratification of target adaptation environments for maize germplasm improvement

Maize breeding research at KARI was organized to improve four classes of germplasm recommended for four distinct production environments:

- The 600-series hybrid for the highland tropics.
- The 500-series hybrid for the midaltitude regions.
- The dryland composite varieties for semiarid areas.
- The coastal composites for the lowland tropics.

The adequacy of this scheme for maize improvement research was evaluated using the database and spatial framework. The resulting analysis of survey information on farmers' choice of varieties, farmers' planting strategies, and time to maize maturity showed a significant deviation from the maize research program's varietal recommendations for target adaptation zones.

Simply stated, the analysis revealed that the original stratification of maize production environments had to be refined. New boundary conditions were delineated to distinguish new zones with different germplasm requirements. For example, a transitional zone between the mid- and high-altitude zones, which comprised about 40% of the area sown to maize in Kenya, was found to require maize germplasm of a different maturity class than the germplasm required in typical mid- and high-altitude climates. Revision of the current classification of agroclimatic zones led to better characterization of maize production environments. The dynamic database and spatial frame developed by the MDBP allow the definition of target adaptation zones to be revised continuously as better information is obtained. Important interactions between factors such as population density, pressure on the land, cropping patterns and intensity, time to maturity, and biotic and abiotic stress factors were found to be the major determinants of farmers' varietal selection and planting regimes.

Priority setting and evaluation of research options

The database and spatial framework were also used to evaluate the relevance of past maize research to problems that farmers considered important and to estimate potential benefits from research on different technologies. Two approaches were used. First, efficiency and nonefficiency indices were constructed using survey data on farmers' practices and statistics on aggregate area and population to evaluate research priorities by agroclimatic zone and production constraint. Zones and production constraints within zones were ranked in priority according to their contribution to efficiency, equity, and sustainability objectives. Farmers' assessments of the most important constraints to maize production and the resulting productivity losses (potential gains from removal) were employed to construct efficiency indices for the constraints identified within adaptation zones. Production constraints were then ranked in order of priority according to the derived efficiency indices.

Resources devoted to research by zone and type of constraint, measured by share in total number of experiments conducted by KARI over the past 12 years, were used to derive an implicit proxy measure of the relative importance or priority assigned to zones and constraints by the maize research program. The priority ranking of the research program was compared with the rankings derived from efficiency and nonefficiency measures and farmers' assessment. The analysis indicated that important adjustments and reorientation in crop improvement and management research emphasis are needed to realize the full production potential in the various zones. Results suggested that KARI needs to reallocate maize research resources to the transitional and midaltitude zones if greater research benefits are to come to farmers.

Measures of economic surplus were used in the second approach to evaluate potential impacts of maize research and set priorities by zone and research program. A model of spatially linked production zones was developed to estimate potential benefits to producers and consumers of maize from different kinds of maize research. The types of research included breeding, crop management research, and research on policies that affect technology dissemination and transfer. Results of this approach were consistent with the index number methods described above, confirming that the potential gains from breeding research are highest in the highland tropics and transitional zones, whereas greater benefits are estimated from crop management research in the relatively marginal environments (semiarid and moist midaltitude).

Monitoring the state of maize technology adoption and transfer

Another important application of the maize database was the assessment of the state of maize technology adoption and diffusion across agroclimatic zones and farmer groups. A comprehensive analysis was done of the determinants of the use and spread of improved seed and inorganic fertilizer, the main elements of first-generation technical change. This analysis updated the only national survey of maize technology adoption in Kenya – conducted in the early 1970s (Gerhart 1975) – and constructed a longitudinal data set to monitor technology adoption by zone, production system, and farmer group (e.g., small-scale, large-scale, male, female). Again, the spatial frame and database will provide a robust system for systematically assessing the evolution in technology transfer and adoption as new data points continue to be added.

Results of the adoption analysis showed both supply and demand forces to be important determinants of the rapidity and extent to which farmers adopted improved maize seed and fertilizer. The most important supply factor was the strategic emphasis on the part of research and delivery systems on the high-potential zones. This emphasis was reflected in the number and range of materials that farmers had to choose from in the highlands and the faster rate at which farmers in highland areas replaced varieties compared with farmers in the low-potential areas. There was almost complete adoption of modern maize varieties in the high-potential zones, whereas adoption of modern varieties remained low and varietal replacement remained slow in the medium- and low-potential areas. Other supply factors that were crucial to the higher adoption of modern maize varieties in high-potential areas included a relatively better developed infrastructure and marketing and seed distribution networks, which made for lower transport and transaction costs. Potential yield increase, financial costs (price), and access to credit, extension, and other farmer support services were the most important factors influencing demand for modern varieties and fertilizer. In the low-potential zones all these effects were weak.

While almost all farmers in the high-potential zones use improved seed, a significant proportion of these farmers apply no or very little fertilizer to their maize crop and hence their improved varieties do not realize their true yield potential. The adoption analyses showed that there is potential to achieve significant gains in maize productivity – current yields could be doubled – through increased use of fertilizers. However, the dominance of low nitrogen fertilizers such as DAP is a major supply factor that must be addressed to provide a balanced supply of nutrients to the maize crop and reduce mining of soil nitrogen.

Assessing the spatial distribution and severity of Striga *infestation in maize*

This application of the integrated survey information and spatial database examined the geographic spread and determinants of the severity of the parasitic weed *Striga* in maize in Kenya. *Striga* infestation in maize was found to be increasing at the margins of the moist transitional zone, Kenya's largest maize-producing agroclimate. In light of this alarming trend, research and extension efforts to contain and prevent *Striga* from spreading into other sensitive zones must be given the highest priority. Given the absence of concrete technologies and research recommendations for controlling *Striga*, farmers' knowledge about the control of *Striga* and *Striga* tolerance in local varieties should be effectively integrated into breeding and crop management research targeting the moist midaltitude zone where *Striga* is a major problem.

Effectiveness of alternative extension strategies

Another application of the spatially referenced database was the analysis of the effectiveness of the T&V approach introduced in Kenya versus the traditional extension methods used previously. Results showed that T&V had been relatively successful in reducing some of the biases of previous extension methods against marginal zones, remote areas, and small-scale and uneducated farmers. However, the bias against female farmers continued under the T&V system.

Strategic Issues and Future Challenges for Maize Research and Policy

Effective utilization, extension, and institutionalization of the database

The MDBP represents an application of spatial analysis and GIS tools to improved planning of agricultural research, technology design, characterization of maize adaptation zones, and targeting of maize technologies. The application of this spatial approach to *ex ante* research evaluation has been extended within KARI to other commodities such as wheat, sorghum, and millet (B. Mills, unpublished paper; B. Mills and D. Karanja, unpublished paper). The maize database and its new zonation scheme were also linked to a crop growth simulation model for maize in Kenya (KARI/IFDC/CIMMYT 1995). There are also plans to replicate the MDBP in other countries and in applications other than research on crops. Nevertheless, many challenges lie ahead if the full potential of spatial techniques and databases is to be exploited fully within and outside KARI. For the database to continue to be developed and utilized by researchers, critical minimum technical and institutional capacities are needed. Some of the challenges that remain crucial to the institutionalization process include:

1. The challenge to maintain the connection in the future between data collection/survey design efforts and the present spatial frame and database, so that information on maize production systems in Kenya is built up systematically.
2. The challenge to utilize and improve the resolution of present coverages and data to address regional farming systems issues.
3. The challenge of utilizing the spatial frame and database to add information from focused surveys and monitor temporal changes in technology diffusion and farming circumstances without the need for repeating expensive national surveys.
4. The challenge of determining whether the objectives we have just described will be served best by integrating the database into line/commodity research programs or by maintaining a centralized database support service role within KARI headquarters.
5. The challenge of moving towards building a well-integrated, single database for all research programs, and of deciding who should manage and coordinate such a database, and how.

Technology design and targeting challenges

The challenges that lie ahead are not, of course, restricted to future uses and institutionalization of the maize database. The research described in this book raises a number of concerns related to future maize technology development and transfer strategies. Each of these concerns needs to be addressed if Kenya is to achieve sustained growth in maize production through the next phase of technical change and transformation.

1. Maize research must shift from a strategy of breeding for yield gains to breeding for pest and disease tolerance. This shift in emphasis is particularly important for the high-potential zones, where adoption of modern varieties is complete, further yield gains are difficult to achieve, and losses to pests and disease damage are substantial. A change in research strategy is also quite important for small-scale farmers who plant hybrids under very poor pest and disease management conditions and fail to benefit from the yield potential of hybrids.
2. Zonal classifications must be revised continuously for a more refined characterization of adaptation zones, design of more appropriate technologies, and more accurate targeting of germplasm and management interventions.
3. More focused maize breeding and crop management research are needed to expand the potential for higher yield gains (or reduced losses) from improved stress tolerance (limited moisture, low soil fertility, and pest and diseases) in relatively marginal environments.
4. The balance between high-input versus low-input strategies for smallholder maize farmers must be evaluated further.
5. Increased investment in research on soil fertility management is needed to address excessive mining of nutrients, particularly nitrogen, from soils planted to maize in areas of increased land pressure and intensive cropping. This problem is made worse by the high cost of commercial fertilizers. Soil fertility management is especially important in high- and medium-potential areas, where the prospective gains from improved fertility management are high because the improved cultivars used in those zones have a high biological response to fertilization. Innovations in alternative sources of nitrogen, more practical methods for enhancing the use of organic fertilizers, nutrient recycling, and reduced soil erosion are all priorities on the research agenda for sustainable maize yields.

Technology support systems and market liberalization

Kenya has relied principally on improved varieties and fertilizer to increase maize productivity. However, effective input delivery systems, supported by extension and credit systems and a favorable economic environment, are critical to such an input-based growth strategy. There is substantial debate over whether a purchased-input strategy is appropriate for African agriculture, particularly in light of the region's limited transportation infrastructure (e.g., see Spencer 1994), high transaction costs in input and output markets, restricted purchasing power of smallholders (see Berry 1993), and the distorted policies and ineffective parastatals through which governments seek to compensate for unfavorable input–output price ratios and underdeveloped, rural capital markets. In many ways, policy as articulated through a range of specialized government institutions, with their roots deep in the colonial period, has been used as a corrective for inadequate infrastructure and incomplete markets. However, this corrective has more often than not created more inefficiencies and drains on fiscal resources than it has eliminated. Structural adjustment and the ensuing market liberalization have shifted the onus back to open markets as the organizational basis for provision of inputs, but infrastructural constraints and high transaction costs to service smallholders remain. How the private sector responds in this new environment will in many ways determine the viability of an input-led productivity strategy. Where input markets remain unprofitable, agricultural research must adjust its technology options to compensate, as technology design becomes the key equilibrating factor in charting alternative strategies to maintain growth in maize productivity.

Adjustments in maize output markets

The Kenya Maize Data Base has been assembled just as Kenya has reached the transition point between a quite highly controlled maize economy – albeit one in which internal prices were not highly distorted in comparison with import parity prices – to an economy that is virtually fully liberalized. Thus the KMDP, in addition to its other roles, provides a baseline to assess the impact of liberalization of the maize economy. What that impact will be has potential implications for research planning.

Three factors are important in assessing the impact of liberalization on the maize market, namely:

1. Price policy before liberalization was based on pan-territorial floor prices set by the National Cereals and Produce Board (NCPB) and based on maize costs of production.
2. Maize yields and supplies are to a significant extent determined by rainfall, which in Kenya is quite variable from year to year.
3. The price band between import and export parity is very wide for maize in Kenya.

Nyoro (1992) estimated an import parity price delivered to Nairobi of KSh 614 per 90 kg bag and an export parity price from Nairobi of KSh 195. Given that there is also a much thinner international market for white maize, with higher price variability (Timmer 1987), the implications are that market price liberalization will provoke a substantial increase in output price variability, some increase in trade, potentially greater variability in per capita consumption, and some shifts in interregional production and trade within Kenya. The large question is what the radical change in relative prices between different agricultural zones within Kenya will do to production patterns, and whether technology development should therefore support a necessary movement toward regional specialization and a search for internal comparative advantage or help to offset that tendency for the major staple food crop in the country.

The elimination of pan-territorial pricing may cause some farmers either to shift out of maize into other crops or to move back to a fully subsistence strategy. The latter is possible as in effect the price bands (the difference between the price at which a farmer can sell maize and the price that a farmer must pay to purchase maize for consumption) faced by individual farmers will also increase as a result of stronger seasonal price movements and stronger price increases in deficit areas (as consumers rather than the NCPB pay the costs of transport from surplus areas). Poorer farmers will not have the option, as they did under controlled prices, to sell maize at harvest for cash and then come back into the market later to meet their consumption needs based on wage or livestock earnings. There is potential for major shifts in cropping patterns and production strategies, with implications for the distribution of agricultural incomes. Monitoring and effective economic evaluation of this dynamic market environment will be needed to determine the best way to adjust research strategies.

Fertilizer markets

Kenyan agriculture, to a very significant extent, has maintained or increased productivity by mining soil capital. The time is fast approaching when Kenyan farmers will have to increase fertilizer use significantly just to maintain existing levels of productivity. Fertilizer will be an essential element of future productivity increases, even with substantial improvement in use of organic sources and improved efficiency in the use of total nutrients. Therefore, what happens to fertilizer use under market liberalization is a critical issue to be considered in devising strategies to improve agricultural productivity in Kenya.

Kenya does not produce fertilizers and in general must contract supplies from international firms. Private firms have operated in the contracting, compounding, and distribution of fertilizers, but until the sector was liberalized in 1989 under the Fertilizer Pricing and Market Reform Program, these firms operated under the restrictions of import licensing, foreign exchange limitations, and price controls. Private sector trade and distribution were particularly depressed in the 1980s because of increased reliance on donor-financed supplies, which reached 63% of total imports in 1987/88 (Kimuyu 1994). The Kenya Grain Growers Cooperative Union (KGGCU, now the Kenya Farmers' Association, KFA) could undercut the private sector but had an ineffective distribution system concentrated in the former White Highlands. This sole agency agreement for donor supplies was canceled in 1985, fertilizer prices were decontrolled in 1990, and foreign exchange and import licensing controls were removed in 1993, leaving a fully liberalized fertilizer sector at the time of the farmer survey undertaken within the MDBP.

There is an assumption in market reform programs that once the liberalization process is fully in place, a competitive market structure will ensure an economically efficient application of fertilizer. Yet the MDBP farmer survey shows that fertilizer use is biased toward large-scale farmers and that small-scale farmers use suboptimal levels of fertilizer even in higher potential areas where fertilizer-responsive varieties are planted and the marginal value to cost ratio is still above three (Argwings-Kodhek *et al.* 1991). Virtually no fertilizer is used in the more marginal agroecosytems. Furthermore, fertilizer use is dominated by a single fertilizer, DAP. All of these findings lead to a number of second-generation issues in the liberalization of the fertilizer sector in Kenya.

The true test of the liberalization process will be whether fertilizer use improves among smallholders producing maize in the very densely populated areas of western Kenya, under a situation of expected deterioration in relative prices, as input/output price ratios will probably be the highest in this region. Output prices will be kept low because of transport costs to Nairobi and proximity to the large-scale, surplus production areas in Trans Nzoia and Uasin Gishu, whereas fertilizer prices will be at their highest because of the high transport costs from Mombasa. Added to this will be a significant increase in price variability, which will also constrain farmers' input use decisions.

Factors to mediate this situation include keeping distribution costs as low as possible, improving the response curve, and improving the efficiency of nutrient use. There will not be much scope to keep distribution costs low, as developing a distribution system founded on supplying small lots to a large number of customers is costly. In fact, because of the purchasing power constraint, there is a need to move to smaller bag sizes. However, adding a significant retailing component to fertilizer distribution networks substantially increases marketing margins and the cost of fertilizer to the farmers. This situation exposes the basic conundrum in marketing systems for small-scale farmers, namely that smallholders need to exploit as many scale economies as possible in distribution (e.g., bulk storage and distribution points in Kisumu). However, these scale economies usually imply reducing marketing agents and competition in the marketing system. There is some evidence of a substantial increase in entry and exit, and price competition, at the retailing end of the market and an increased concentration at the upper end of the market (Kimuyu 1994), which should favor the development of such scale economies. Nevertheless, where investments will be made in bulking and distribution points for fertilizers still remains a question under the market liberalization process.

The largest efficiency gains to be achieved are probably in improved management of the nutrients that are applied. There is significant scope for better targeting of fertilizers in terms of fertilizer type, nutrient balance, nutrient placement, and timing of nutrient application to improve the yield response to fertilizer application. Nevertheless, the question is how to link such agronomic and soils research to better targeted compounding, packaging, and distribution systems, when both farmers and distributors have limited experience with fertilizers other than DAP.

What is apparent is that more information on the fertilizer system is needed to guide the investments and decisions being made by all the actors in this chain. This includes monitoring fertilizer prices, quantities, and marketing channels, as well as information for better targeting of fertilizer products. Integrating the KMDP with FURP is a start in that direction, but it requires better links to a monitoring system. As of now, there is little capacity to monitor the fertilizer system in Kenya, although the Fertilizer Inputs Branch of the Ministry of Agriculture has the mandate to do so. An understanding of how Kenya responds to liberalization of the fertilizer market will have significant implications for such countries as Uganda, Rwanda, and Burundi, where these issues are even more binding.

Maize seed systems

Seed release and certification regulations have been a mechanism to control quality within the maize seed distribution system. Such regulations have also been a mechanism to control competition and capture rents. The seed sector is also part of the liberalization process, but reform here has been slower than in the fertilizer sector, with the impact of liberalization still difficult to determine.

The maize seed market in Kenya is quite highly segmented. The higher potential agroecological zones depend on hybrids, with areas sown to different hybrids segmented by germplasm maturity class. The more semiarid areas depend on OPVs, and the coastal area on either OPVs or a recently developed hybrid. Further segmentation in the market is possible, as was explained above for the transitional zone. Kenya's maize breeding program has been relatively successful in developing first-generation varieties for most of these market niches. Varietal development and seed multiplication and distribution have remained very much a Kenyan activity, with strict requirements on varietal release and multiplication. The monopolistic position of the KSC in the seed market has in fact resulted in a slow deterioration of that company's ability to supply seed of the right type, in sufficient quantity, at the time required.

Deregulation of the seed system to date has been based on KARI's ability to sell foundation seed on a royalty basis to any established seed organization, undercutting the monopoly position of KSC. There is the intention to open the maize seed sector up to multinational seed companies, most of which have a base in southern Africa. How effectively varieties from these companies will perform in Kenya remains to be seen. Certainly some of the multinationals' hybrids are well ahead of current materials in having maize streak virus resistance, but competing in Kenya's niche markets will probably require some adaptive breeding, with the potential of very limited market size. In addition, it is doubtful whether multinationals will move into the OPV market, which is probably greatest in the semiarid zone where OPVs have a clear advantage. It is probable that the multinationals will try to compete in those niches where existing varieties have some competitive potential. What is clear is that while competitive pressure within the Kenya seed market is needed, the private sector will not meet the requirements of all of Kenya's maize farmers, who will require a continuing research role by the public sector, namely KARI. The eventual division of labor between the private and public sectors will depend in large part on the performance of proprietary germplasm from the multinationals. What might be most attractive is a cooperative agreement between KARI and the multinationals.

Conclusion

The Kenya maize sector is at something of a juncture. Although Kenya can justly be proud of past achievements in the maize sector, it is necessary to understand the nature of that success, and it is even more important to recognize that the critical need for sustained, future increases in maize productivity will require some shifting of gears, led in part by the market liberalization process.

There will be needed changes in breeding strategy, which the public sector can lead, even with increased competition from multinational seed companies. Probably even more important will be the adjustments needed in the fertilizer sector and the development of critical linkages between the public and private sectors.

Both the private and public sectors face the same structural constraints of inadequate infrastructure, significant diversity in technology requirements within the country, and the high transaction costs inherent in smallholder market systems. And just as the public sector faces inadequate public sector budgets, the private sector faces very major capital and credit constraints, especially at the bottom end of the distribution channels.

Future development of the Kenyan maize sector will be a joint private and public sector collaboration. Increasing the information flow into this process is critical. The MDBP offers a new base on which to evaluate the research plans and to fine tune the market strategies that will lead to the next generation of productivity gains in maize cropping systems in Kenya.

References

Argwings-Kodhek, G., Omamo, W., Karin, F., and Okuma, G. (1991) *Fertilizer Marketing Study.* Policy Analysis Matrix, Egerton University, Njoro, Kenya. Mimeo.

Berry, S. (1993) *No Condition is Permanent: The Social Dynamics of Agrarian Change in Sub-Saharan Africa.* University of Wisconsin, Madison, Wisconsin.

Gerhart, J. (1975) *The Diffusion of Hybrid Maize in Western Kenya.* Research Report for CIMMYT. International Maize and Wheat Improvement Center (CIMMYT), Mexico City. Mimeo.

KARI/IFDC/CIMMYT (1995) *An Integrated Approach to Assessing Soil Fertility and Climatic Interactions in Pilot Maize-producing Areas of Kenya.* Project Proposal Document by International Fertilizer Development Center (IFDC) and CIMMYT. Kenya Agricultural Research Institute (KARI), Nairobi, Kenya.

Kimuyu, P.K. (1994) *Evaluation of the USAID/Kenya Fertilizer Pricing and Marketing Reform Program.* Final Report, USAID, September 1994. Nairobi, Kenya.

Nyoro, J.K. (1992) Competitiveness of maize production systems in Kenya. In: Nguyo, W. (ed.) *Proceedings of the Conference on Maize Supply and Marketing under Market Liberalization.* Policy Analysis Matrix, Egerton University, Njoro, Kenya.

Spencer, D. (1994) *Infrastructure and Technology Constraints to Agricultural Development in the Humid and Subhumid Tropics of Africa.* EPTD Discussion Paper No. 4. International Food Policy Research Institute (IFPRI), Washington, DC.

Timmer, C.P. (1987) *The Corn Economy of Indonesia.* Cornell University, Ithaca, New York.

Appendix A
Computation of Thermal Time

The sum of an estimate of mean daily temperature between a base temperature level (T_b) at which the rate of maize development is zero and an optimal temperature (T_o) at which development rate is maximal is used to estimate heat units or thermal time (Bonhomme *et al.* 1994) (see Chapter 2). The general form for calculating thermal time in °C days is:

$$\mathrm{TM} = \sum_{i}^{n} \{[(T_x + T_n/2)] - T_b\}, \tag{A.1}$$

where: TM = thermal time, in (°C days; T_x = the daily maximum temperature, in °C; T_n = the daily minimum temperature, in °C; T_b = the base temperature, in °C; and n = the duration under investigation, in days.

When (T_x+T_n) is below T_b, it is set equal to T_b.

An optimal temperature level of 30°C (T_o = 30) was determined suitable for tropical maize development (Abasi *et al.* 1985). When T_x exceeds 30°C, the difference is subtracted from 30 before mean temperature is computed (Bonhomme *et al.* 1994). The base temperature level for tropical maize was estimated at 7°C, compared with 6°C for temperate maize (G. Edmeades, pers. comm.). Average thermal time to maturity for maize germplasm grown in various agroclimates in Kenya is computed in Table A.1 using T_o = 30°C and T_b = 7°C.

Table A.1. Average thermal time to maturity in various maize agroclimatic zones, Kenya.

Agroclimatic zone	Mean daily maximum temperature, T_x (°C)	Mean daily minimum temperature, T_n (°C)	Mean daily heat units (°C)	Average days to maturity	Thermal time (°C days)
Lowland tropics	29.4	20.0	17.7	120	2124
Semiarid zone	28.6	16.4	15.5	114	1767
Moist midaltitude zone	28.3	15.9	15.1	163	2461
Dry transitional zone	25.3	14.0	12.7	144	1829
Moist transitional zone	23.3	13.4	11.4	181	2063
Highland tropics	23.0	10.0	9.7	213	2066

References

Abasi, L., Fakorede, M.B., and Alofe, C.O. (1985) Comparison of heat units and calendar days for predicting silking dates in maize in a tropical rainforest location. *Maydica* 30, 15–30.

Bonhomme, R., Derieux, M., and Edmeades, G. (1994) Flowering of diverse maize cultivars in relation to temperature and photoperiod in multilocation field trials. *Crop Science* 34, 156–164.

Appendix B
Measuring the Benefits of Research over Time with Multiple, Spatially Linked Production Zones

The main components of the spatial multimarket model used in Chapter 6 to measure the impact of maize research are presented below.

Shifts in Supply Induced by Research

Successful research is assumed to induce a parallel downward shift in the commodity supply curve. For each zone, the cost reduction expected from research, K_{it}, is simply calculated for every period as the product of the probability of exceeding the dissemination threshold, the expected net yield gain (conditional upon the dissemination threshold being exceeded), the expected adoption rate for the period, and the initial unit price of the commodity divided by the supply elasticity for the zone. Thus

$$K_{it}, = Pr(K_i > K_i^a)E[K_i | K_i > K_i^a]A_{it}P_{i0}/\epsilon_i, \tag{B.1}$$

where A_{it} is the adoption rate in zone i at time t as determined by the trapezoid adoption profile and parameters in Fig. 6.3, P_{i0} is the initial commodity price in the zone, and ε_i is the supply elasticity in the zone.
The initial linear supply curve for a given zone is specified as:

$$Qs_{i0} = \alpha_{i0} + B_i P_{i0}, \tag{B.2}$$

where Qs_{i0} is the initial quantity supplied in zone i, α_{i0} is the initial supply intercept in zone i, and B^i is the fixed supply slope parameter in zone i. The initial supply intercept and fixed slope parameter are easily calculated from the initial supply, price, and elasticity for the zone.

The unit cost reduction for a specific point in time t then translates into a research-induced change in the quantity of the commodity supplied, as follows:

$$Q^r s_{it} = \alpha^R{}_{it} + B_i P_{it}, \tag{B.3}$$

where $\alpha^R_{it} = \alpha_{i0} + K_{it}B_i$ is the intercept of the 'with research' supply curve for zone i in period t.

Shifts in Demand as a Result of Population Growth

The shift in demand for maize resulting from population increases can also be included in the model. Let the linear demand curve for zone i in period t be expressed as:

$$Qd_{it}=\gamma_{it} + \delta_i P_{it}, \qquad \text{(B.4)}$$

where Qd_{it} is the quantity demanded in zone i at time t, γ_{it} is the demand intercept in zone i at time t, and δ_i is the fixed demand slope parameter for zone i. Again, the zone-specific demand intercept and slope parameters are calculated from the initial demand, price, and elasticity. Population growth will result in an increase in the intercept term of the demand function in period t + 1 of:

$$\gamma_{it}+1=\gamma_{it}+\pi_i Qd_{it}, \qquad \text{(B.5)}$$

where π_i is the population growth rate for zone i.

Constant Price Wedges between Zones

Prices in zone i for period t, P_{it}, are specified in terms of the Nairobi price P_{nt}, net of a price wedge, T_i, which reflects the transactions costs of shipping surplus maize to (or from) Nairobi. T_i is constant in real terms over time. Thus:

$$P_{it} = P_{nt} - T_i. \qquad \text{(B.6)}$$

Market Clearing Conditions

For each period, equilibrium quantities and prices are determined for two scenarios ('with research' and 'without research') through the respective market clearing conditions:

$$\sum_i Qs_{it} = \sum_i Qd_{it} \text{ and } \sum_i Q^R_{it} = \sum_i Q^R d_{it} \qquad \text{(B.7)}$$

If equilibrium quantities differ in the two scenarios, zonal prices will also differ in the 'without research' scenario (P_{it}) and the 'with research' (P^R_{it}) scenario.

Measures of Producer and Consumer Surplus

Changes in producer and consumer surplus (ΔPS and ΔCS) as a result of research are easily calculated from equilibrium quantities and prices for the 'with research' and 'without research' scenarios. The change in producer surplus in zone i at time t is calculated as:

$$\Delta PS_{it} = (K_{it} + P^{R}_{it} - P_{it})[Q_{it} + 0.5(Q^{r}_{it} - Q_{it})]. \qquad (B.8)$$

The corresponding change in consumer surplus is:

$$\Delta CS_{it} = (P_{it} - P^{R}_{it})[Q_{it} + 0.5(Q^{r}_{it} - Q_{it})]. \qquad (B.9)$$

The present values for the stream of changes in producer and consumer surplus (VPS and VCS) over the 30-year research planning horizon for each zone are:

$$VPS_{i} = \sum_{t=0}^{30} \Delta PS_{it} / (1 + r)^{t} \text{ and } VCS_{i} = \sum_{t=0}^{30} \Delta CS_{it} / (1 + r)^{t}, \qquad (B.10)$$

where r is the real discount rate for the use of public sector financial resources. In this study, the real discount rate is assumed to be 5%, based on the interest rate of government loans to the agricultural sector. Surplus changes can be added across zones to assess the total impact (within and across zones) of spatially targeted research.

Appendix C
The Potential for Generation and Adoption of Technologies by Zone and Major Research Theme

The research priority setting exercise described in Chapter 6 was partly based on opinions obtained from a working group of key maize program scientists from different disciplines and regions of Kenya. These scientists reviewed three major maize research themes (defined in Chapter 6): varietal development research, crop management research, and technology environment research. The working group reviewed the major constraints to maize production in each agroclimatic zone, identified the key problems to be addressed by the three research themes, and identified the research impacts that could be made in each zone by addressing these constraints. The assumptions that guided the group are presented here in Tables C.1, C.2, and C.3. Note that the working group restricted its evaluation of technology environment research to activities related to the dissemination of technology. For a full discussion of the reasons for this choice, see Chapter 6.

Table C.1. The potential for generation and adoption of technologies in Kenya through varietal development research, by zone.

Zone (facts and assumptions)	Constraints to varietal development (ranked)	Potential for generation and adoption of technologies
Lowland tropics • Yield potential in zone is higher for composites than for hybrids • Grain texture and storability must be taken into account in varietal development • Farm yields average less than 1 t ha^{-1} whereas yields on research stations average 4–5 t ha^{-1} • A large portion of the yield gap is attributed to poor infrastructure for delivery of seeds and inputs • Farmer, researcher, and extension links are weak	• Long time to maturity of current varieties • Maize streak virus • Poor grain texture in improved varieties • Low yield stability under variable rainfall	***Technology generation*** • Minimum expected yield increase = 4.2% • Most likely expected yield increase = 12.5% • Maximum expected yield increase = 16.6% • Yield increases will come primarily from composite varieties with few additional production costs incurred (net yield gains approximately equal gross gains) • The minimum net yield increase necessary for varieties to be disseminated to farmers = 10% • Probability of dissemination = 67% • Conditional expected net yield increase = 12.57% ***Technology adoption*** Given the relatively low expected yield increases, poor delivery systems for inputs, and weak links between research, extension, and farmers, adoption will occur slowly, with: • A ten-year research lag • Ten years to maximum adoption after release • A maximum adoption rate of 25% of maize growers • Abandonment of varieties commences 15 years after maximum adoption • Ten years to complete abandonment
Dry midaltitude zone • Potential for development of varieties is limited, given the stability of current germplasm to the variable environment	• Drought and heat stress • Low yields when rainfall is not limiting • Poor soil fertility	***Technology generation*** • Minimum expected yield increase = 5% • Most likely expected yield increase = 7.5% • Maximum expected yield increase = 10% • Few additional production costs will be incurred from adoption of improved varieties (net yield gains approximately equal gross gains) • The minimum net yield increase necessary for varieties to be disseminated to farmers = 10% • Given the low expected net yield increases, the probability of dissemination is close to zero

Table C.1. (cont'd.)

Zone (facts and assumptions)	Constraints to varietal development (ranked)	Potential for generation and adoption of technologies
Moist midaltitude zone • Farmers will need to switch from local to hybrid varieties • The seed distribution system must be functional	Current varieties are: • Slightly too late, especially for the two-season production system in the western part of the zone • Low tolerance to pests and diseases (*Striga* in the western part of the zone and streak virus in the eastern part)	***Technology generation*** • Minimum expected yield increase = 5% • Most likely expected yield increase = 15% • Maximum expected yield increase = 20% • The costs associated with moving to hybrid seed will equal one-fourth of the cost of yield increases, leaving minimum expected net yield increases of 3.75%; most likely expected net yield increases of 11.5%; maximum expected net yield increases of 15% • The minimum net yield increase necessary for varieties to be disseminated to farmers = 11.5% • Probability of dissemination = 31% • Conditional expected net yield increase = 12.53% ***Technology adoption*** Given the expected yield increases, moderate delivery systems for inputs, and functional links between research, extension, and farmers, adoption will occur relatively quickly, with: • A ten-year research lag • Seven years to maximum adoption after release • A maximum adoption rate of 60% of maize growers • Abandonment of varieties commences 15 years after maximum adoption • Seven years to complete abandonment
Dry transitional zone • Currently both 500-series hybrids and Katumani varieties are grown • The availability of germplasm (from CIMMYT, for instance) suggests quick results from adaptation and screening are possible	• Current varieties are not well suited to the length of growing period in the zone	***Technology generation*** • Minimum expected yield increase = 10% • Most likely expected yield increase = 20% • Maximum expected yield increase = 30% • Few additional production costs will be incurred from adoption of improved varieties (net yield gains approximately equal gross gains) • The minimum net yield increase necessary for varieties to be disseminated to farmers = 10% • Probability of dissemination = 100% • Conditional expected net yield increase = 20% ***Technology adoption*** Given the relatively high expected yield increases, adoption will occur relatively quickly, with: • A ten-year research lag • Five years to maximum adoption after release • A maximum adoption rate of 60% of maize growers • Abandonment of varieties 15 years after maximum adoption • Five years to complete abandonment

<table>
<tr>
<td>Moist transitional zone
• The 600-series hybrids have a high yield potential in the first rains, but their long time to maturity interferes with the second-season planting
• The current adoption level of improved varieties in the zone is high</td>
<td>• Lack of suitable varieties for the current two-season cropping sequence
• Pests and diseases, particularly stalk borer, smut rot, and streak virus, are prevalent in the zone</td>
<td>Technology generation
• A variety better suited for the first cropping season would have a major impact on yields in both cropping seasons. Minimum expected yield increase = 15%; most likely expected yield increase = 30%; maximum expected yield increase = 50%
• Few additional production costs will be incurred from adoption of the improved variety (net yield gains approximately equal gross gains)
• The minimum net yield increase necessary for the variety to be disseminated to farmers = 10%
• Probability of dissemination = 100%.
• Conditional expected net yield increase = 31.67%
Technology adoption
Given the high expected yield increases and good seed delivery system in the zone, adoption will occur relatively quickly, with:
• A ten-year research lag
• Six years to maximum adoption after release
• A maximum adoption rate of 80% of maize growers
• Abandonment of varieties 20 years after maximum adoption
• Six years to complete abandonment</td>
</tr>
<tr>
<td>Highland tropics
• Previous efforts to develop new varieties have focused on this zone
• The current level of adoption of improved varieties in the zone is high
• Maintenance research will play an important role in preventing resistance in current varieties from breaking down</td>
<td>• Prevalence of pests and head smut
• Low levels of soil fertility</td>
<td>Technology generation
• Minimum expected net yield increase = 5%
• Most likely expected net yield increase = 10%
• Maximum expected net yield increase = 15%
• Given the current level of use of improved varieties, the minimum net yield increase necessary for a new variety to be disseminated to farmers = 5%
• Probability of dissemination = 100%
• Conditional expected net yield increase = 10%
Technology adoption
Given the good seed delivery system in the zone, adoption will occur relatively quickly, with:
• A ten-year research lag
• Five years to maximum adoption after release
• A maximum adoption rate of 90% of maize growers
• Abandonment of varieties 12 years after maximum adoption
• Five years to complete abandonment</td>
</tr>
</table>

Table C.2. The potential for generation and adoption of technologies in Kenya through crop management research, by zone.

Zone (facts and assumptions)	Crop management constraints (ranked)	Potential for generation and adoption of technologies
Lowland tropics • The zone has a large yield gap because of current low-input management practices at the farm level • Improved management systems will need to intensify the use of external inputs	• Weed control problems resulting from high temperature and rainfall • Sandy soils have poor fertility • Poor stand establishment (particularly when maize is intercropped)	***Technology generation*** • Minimum expected net yield increase = 30% • Most likely expected net yield increase = 50% • Maximum expected net yield increase = 70% • The additional inputs associated with these yield increases would equal half of the cost of yield increases, leaving a minimum expected net yield increase of 15%; most likely expected net yield increase of 25%; maximum expected net yield increase of 35% • The reliance on external inputs will raise the net yield increase necessary for a technology to reach a dissemination phase to 20% • Probability of dissemination = 88% • Conditional expected net yield increase = 25.95% ***Technology adoption*** The heavy reliance on external inputs for crop management technologies suggests that the rate and extent of adoption will be low, with: • A five-year research lag • Twenty years to maximum adoption after release • A maximum adoption rate of 10% of maize growers • No significant abandonment of the improved techniques is expected in the 30-year research horizon
Dry midaltitude zone • Timely preparation of land can result in significant yield increases • Yield increases will be much greater if accompanied by more effective soil fertility management strategies	• Poor seedbed preparation • Low levels of soil fertility • Poor weed control • Disease and pest pressure	***Technology generation*** • Improved crop management strategies to address these constraints can be expected to result in a minimum yield increase of 20%; most likely yield increase of 35%; and maximum yield increase of 50% • 35% of the value of yield increases is embodied in the increased cost of inputs • Expected minimum net yield increase = 13%; most likely net yield increase = 22.75%; maximum net yield increase = 32.5% • The reliance on external inputs suggests net yield increases of at least 15% are necessary for technologies to be disseminated to farmers • Probability of dissemination = 98% • Conditional expected net yield increase = 22.93% ***Technology adoption*** The heavy reliance on external inputs for crop management technologies suggests that the rate and extent of adoption will be moderate, with: • A six-year research lag. • Twelve years to maximum adoption after release • Maximum adoption by 40% of the farmers • No significant abandonment of the technologies is expected in the 30-year research horizon

Moist midaltitude zone • Farmers are willing to invest in more intensive soil fertility management systems • Input delivery systems must be strengthened	• Low soil fertility • *Striga* infestation • Poor weed control • High pest pressure	***Technology generation*** • Improved crop management strategies to address these constraints could result in a significant minimum yield increase of 30%; most likely yield increase of 45%; maximum yield increase of 60% • 40% of the value of yield increases will be embodied in increased cost of inputs (i.e., fertilizer) • Expected minimum net yield increase = 21% • Most likely net yield increase = 28% • Maximum net yield increase = 42% • The reliance on external inputs suggest net yield increases of at least 20% are necessary for technologies to be disseminated to farmers • Probability of dissemination = 100% • Conditional expected net yield increase = 30.33% ***Technology adoption*** The heavy reliance on external inputs for crop management technologies suggests that the rate and extent of adoption will be very low, with: • A five-year research lag • Sixteen years to maximum adoption after release • Maximum adoption by 20% of farmers • No significant abandonment of the technologies is expected in the 30-year research horizon
Dry transitional zone • Soil fertility is a binding constraint • Many farmers use inorganic fertilizer, so research should focus on strategies for managing organic/inorganic matter	• Low soil fertility (specifically organic matter) • Poor moisture management strategies • Poor weed management • Lack of effective crop protection techniques	***Technology generation*** • A technology package that addresses these constraints can significantly increase yields, offering a minimum yield increase of 20%; most likely yield increase of 35%; maximum yield increase of 50% • 35% of the value of yield increases will be embodied in increased cost of inputs • Expected minimum net yield increase = 13% • Most likely net yield increase = 22.75% • Maximum net yield increase = 32.5% • The moderate reliance on external inputs suggests net yield increases of at least 15% are necessary for technologies to be disseminated to farmers • Probability of dissemination = 98% • Conditional expected net yield increase = 22.93% ***Technology adoption*** The moderate reliance on external inputs also suggests that adoption will grow and expand at a relatively fast rate, with: • A five-year research lag • Ten years to maximum adoption after release • Maximum adoption by 50% of farmers • No significant abandonment of the technologies is expected in the 30-year research horizon

Table C.2. (cont'd.)

Zone (facts and assumptions)	Crop management constraints (ranked)	Potential for generation and adoption of technologies
Moist transitional zone • The variety of related crop management constraints in the zone suggests that the potential for generating an improved crop management package is relatively low	• Prevalence of disease and pests • Low soil fertility • Late planting • Poor weed control • Moisture stress during the short rains • Lack of effective intercropping strategies	***Technology generation*** • Minimum net expected yield increase = 5% • Most likely net expected yield increase =10% • Maximum net expected yield increase = 20% • Moderate reliance on external inputs suggests that net yield increases of at least 15% are necessary for technologies to be disseminated to farmers • Probability of dissemination = 17% • Conditional expected net yield increase = 16.46% ***Technology adoption*** The moderate reliance on external inputs suggests that adoption will occur relatively quickly if the 15% net yield increase threshold is exceeded, with: • A five-year research lag. • Twelve years to maximum adoption after release • Maximum adoption by 50% of farmers • No significant abandonment of the technologies is expected in the 30-year research horizon
Highland tropics • Farmers used fertilizers and high management levels from the 1960s to the 1980s • Dramatic increases in fertilizer prices reduced or eliminated use of these inputs • Given current input market policies, the potential for improving crop management strategies is limited, and research must focus on more cost-effective use of inputs • However, the input delivery system is relatively good in the zone, and farmers will readily adopt these more cost-effective management systems	• Low soil fertility • Erratic onset of rains interferes with land preparation and time of planting • Poor weed control strategies	***Technology generation*** • Minimum expected yield increase = 5% • Most likely expected yield increase = 10% • Maximum expected yield increase = 15% • 25% of the value of yield increases will be embodied in increased cost of inputs • Expected minimum net yield increase = 3.75% • Most likely net yield increase = 7.5% • Maximum net yield increase = 11.25% • Only a 10% net yield increase would be needed for farmers to adopt more cost-effective management strategies • Probability of dissemination = 6% • Conditional expected net yield increase = 10.37% ***Technology adoption*** The rate and extent of adoption will occur quickly if the 10% net yield increase threshold is exceeded, with: • An eight-year research lag • Ten years to maximum adoption after release • Maximum adoption by 60% of farmers • No significant abandonment of the techniques is expected

Table C.3. The potential for generation and adoption of technologies in Kenya through technology dissemination research, by zone.

Much discussion ensued in the working group on the components of technology dissemination research. Three major components of this kind of research were considered particularly relevant for the KARI maize program:

1 Characterizing farmers' maize production systems to improve the relevance of technologies generated by researchers.
2 Improving links between farmers and extension to increase the dissemination and adoption of technologies.
3 Conducting studies of the broader technology environment and policy to advise on policy options for improving maize productivity.

The primary output of these components is a better process for generating and disseminating technologies. Measuring the impact of these components in terms of net yield increases is difficult, however, because researchers must conceptualize how changes in the institutional processes of generating and disseminating technology will affect research output and adoption by farmers. In addition, this research thrust essentially complements the other research thrusts. The framework within which technology dissemination efforts translate into research benefits is thus arguably different than for the other research themes.To maintain compatibility across themes, the working group used the same structured process for all of them; to minimize overlap between benefits arising from technology dissemination research and from the other research themes, the working group focused on the potential for increasing the flow of available technologies to farmers.

Zone	Technology dissemination (facts and assumptions)	Potential for generation and adoption of technologies
Lowland tropics	• There is a large gap between maize yields obtained on experiment stations and in farmers' fields. • Available technologies can significantly close this yield gap and result in large net yield increases for farmers	***Technology generation*** • Minimum net expected yield increase = 25% • Most likely net expected yield increase = 30% • Maximum net expected yield increase = 35% • Only a 7% net yield increase would be needed for farmers to adopt currently available technologies • Probability of dissemination = 100% • Conditional expected net yield increase = 30% ***Technology adoption*** The process for improving the institutional structures necessary to disseminate available technologies is lengthy, with: • A ten-year lag before improved links between farmers, researchers, and extension are felt at the farm level • Currently available technologies are then expected to take 30 years to reach maximum adoption • Only 10% of farmers will eventually adopt these technologies • No significant abandonment of these technologies will occur in the 30-year research planning horizon

Table C.3. (cont'd.)

Zone	Technology dissemination (facts and assumptions)	Potential for generation and adoption of technologies
Dry midaltitude zone	• There is a large gap between maize yields obtained on experiment stations and in farmers' fields • Available technologies can significantly close this yield gap and result in large net yield increases for farmers	***Technology generation*** • Minimum net expected yield increase = 10% • Most likely net expected yield increase = 20% • Maximum net expected yield increase = 30% • Only a 7% net yield increase would be needed for farmers to adopt currently available technologies • Probability of dissemination = 100% • Conditional expected net yield increase = 20% ***Technology adoption*** The process for improving the institutional structures necessary to disseminate available technologies is lengthy, with: • A ten-year lag before improved links between farmers, researchers, and extension are felt at the farm level • Currently available technologies are then expected to take ten years to reach maximum adoption • 20% of farmers will eventually adopt these technologies • No significant abandonment of these technologies will occur in the 30-year research planning horizon
Moist midaltitude zone	• There is a large gap in maize yields obtained on experiment stations and in farmers' fields • Available technologies can significantly close this yield gap and result in large net yield increases for farmers	***Technology generation*** • Minimum net expected yield increase = 10% • Most likely net expected yield increase = 20% • Maximum net expected yield increase = 30% • Only a 7% net yield increase would be needed for farmers to adopt currently available technologies • Probability of dissemination = 100% • Conditional expected net yield increase = 20% ***Technology adoption*** The process for improving the institutional structures necessary to disseminate available technologies is lengthy, with: • A ten-year lag before improved links between farmers, researchers, and extension are felt at the farm level • Currently available technologies are then expected to take 10 years to reach maximum adoption • 30% of farmers will eventually adopt these technologies • No significant abandonment of these technologies will occur in the 30-year research planning horizon

Dry transitional zone	• There is a large gap between maize yields obtained on experiment stations and in farmers' fields • Available technologies can significantly close this yield gap and result in large net yield increases for farmers	***Technology generation*** • Minimum net expected yield increase = 20% • Most likely net expected yield increase = 25% • Maximum net expected yield increase = 30% • Only a 7% net yield increase would be needed for farmers to adopt currently available technologies • Probability of dissemination = 100% • Conditional expected net yield increase = 25% ***Technology adoption*** The process for improving the institutional structures necessary to disseminate available technologies is lengthy, with: • A ten-year lag before improved links between farmers, researchers, and extension are felt at the farm level • Currently available technologies are then expected to take ten years to reach maximum adoption • 20% of farmers will eventually adopt these technologies • No significant abandonment of these technologies will occur in the 30-year research planning horizon
Moist transitional zone	• The gap between yields obtained on experiment stations and in farmers' fields is smaller than in other zones • Institutional links and structures for disseminating maize research results are already strong	***Technology generation*** • Minimum net expected yield increase = 10% • Most likely net expected yield increase = 15% • Maximum net expected yield increase = 20% • Only a 7% net yield increase would be needed for farmers to adopt currently available technologies • Probability of dissemination = 100% • Conditional expected net yield increase = 15% ***Technology adoption*** The process for improving the institutional structures necessary to disseminate available technologies is lengthy, with: • A ten-year lag before improved links between farmers, researchers, and extension are felt at the farm level • Currently available technologies are then expected to take 10 years to reach maximum adoption • 20% of farmers will eventually adopt these technologies • No significant abandonment of these technologies will occur in the 30-year research planning horizon

Table C.3. (cont'd.)

Zone	Technology dissemination (facts and assumptions)	Potential for generation and adoption of technologies
Highland tropics	• The gap between yields obtained on experiment stations and in farmers' fields is smaller than in other zones • Institutional links and structures for disseminating maize research results are already strong	***Technology generation*** • Minimum net expected yield increase = 5% • Most likely net expected yield increase = 10% • Maximum net expected yield increase = 20% • Only a 7% net yield increase would be needed for farmers to adopt currently available technologies • Probability of dissemination = 95% • Conditional expected net yield increase = 11.97% ***Technology adoption*** The process for improving the institutional structures necessary to disseminate available technologies is lengthy, with: • A ten-year lag before improved links between farmers, researchers, and extension are felt at the farm level • Currently available technologies are then expected to take ten years to reach maximum adoption • 40% of farmers will eventually adopt these technologies • No significant abandonment of these technologies will occur in the 30-year research planning horizon

Appendix D
Improved Maize Varieties Released in Kenya

Table D.1. Improved maize varieties and hybrids released in Kenya.

Variety/hybrid	Year released	Age in 1992	Percentage of area planted to variety/hybrid in 1992/93[a]	Weighted average age of variety/hybrid
Used in 1992/93				
Coast Composite	1974	18	0.95	0.17
Katumani Composite B	1968	24	5.33	1.28
Makueni Composite	1989	3	0.24	0.01
Hybrid Pwani	1989	3	0.59	0.02
Hybrid H511	1968	24	7.23	1.17
Hybrid H512	1970	22	3.67	0.81
Hybrid H632	1965	27	0.36	0.10
Hybrid H622	1965	27	2.13	0.58
Hybrid H613 D	1986	6	2.13	0.13
Hybrid H614 D	1986	6	41.83	2.51
Hybrid H525	1981	11	22.87	2.50
Hybrid H626	1989	3	12.80	0.38
Not used in 1992				
Kitale Synthetic II	1961	31	0.0	na
Katumani Synthetic II	1963	29	0.0	na
Katumani Composite A	1966	26	0.0	na
Hybrid H611	1964	28	0.0	na
Hybrid H621	1964	28	0.0	na
Hybrid H631	1964	28	0.0	na
Hybrid H611 C	1971	21	0.0	na
Hybrid H612 C	1966	26	0.0	na
Hybrid H613 C	1972	20	0.0	na
Hybrid H614 C	1976	16	0.0	na
Hybrid H612 D	1986	6	0.0	na
			100.0	10.00

Source: Kenya Agricultural Research Institute, Maize Data Base records of crop improvement experiments, and farmer survey (1992/93).
na = Data not available.
[a] These figures represent the percentage share of area sown to improved varieties and hybrids only, which occupy 73% of the area planted to maize in Kenya.

Appendix E
Recent Trends in the Production and Marketing of Maize Seed in Kenya, 1980–1993

Table E.1. Annual production and sales of maize seed by Kenya Seed Company (1980–1993).

Year	Commercial seed production (000 t)	Seed sales (000 t)	Retail seed price (KSh kg^{-1})	Retail grain price (KSh kg^{-1})	Ratio of seed:grain price	Number of seed packages (000s)		
						25-kg package	10-kg package	2-kg package
1980	15.7	12.2	4.4	na	na	58.9	1072	0
1981	22.5	13.5	5.5	na	na	56.9	1206	0
1982	16.4	13.0	5.5	1.9	2.9	78.0	1106	0
1983	14.1	15.5	7.2	2.2	3.3	88.1	1285	216
1984	14.2	18.9	7.2	2.6	2.8	116.9	1527	339
1985	22.7	20.7	7.2	3.1	2.3	144.9	1644	319
1986	25.1	20.4	8.7	3.1	2.8	163.6	1553	408
1987	23.7	21.8	8.7	3.1	2.8	198.5	1594	468
1988	27.6	22.1	10.4	3.2	3.3	108.6	1831	517
1989	22.2	17.2	10.4	3.4	3.1	103.1	1328	687
1990	15.8	19.6	12.5	3.9	3.3	145.1	1479	605
1991	18.8	21.6	17.0	5.0	3.4	132.7	1515	157
1992	na	na	17.0	6.7	2.5	na	na	na
1993	na	na	28.0	10.0	2.8	na	na	na
Average					2.9			

na = Data not available.

Table E.2. Local seed sales by variety (Kenya Seed Company, 1987–1991).

Variety/hybrid	1987	1988	1989	1990	1991	Average
H613	12.3	3.9	1.1	0.1	0.0	
H614	40.8	52.7	52.0	57.3	57.5	
H632	1.7	0.5	0.8	0.4	0.0	
H622	4.3	3.4	1.7	2.8	2.9	
H625	27.9	29.0	33.9	21.7	9.2	
H626	0.0	0.3	0.8	5.7	15.7	
Total, late-maturing	87.0	89.8	90.3	88.0	85.3	88.1
H511	5.3	5.0	3.4	4.5	3.4	
H512	4.7	3.0	4.3	5.6	5.3	
Total, medium-maturing	10.0	8.0	7.7	10.1	8.7	8.9
Other hybrids[a]	0.6	0.45	0.75	0.9	2.2	1.0
Total hybrids	97.6	98.25	98.75	99.0	96.2	98.0
Improved OPVs[b]	2.4	1.75	1.25	1.0	3.8	2.0

Source: Kenya Seed Company, unpublished records, Kitale, Kenya (1994).

[a] Mainly Pwani hybrid targeted for coastal areas.

[b] Mainly Katumani, Coast, and Makueni Composites.

Appendix F
FURP Trials (Type 1 and Type 2 Experiments)

Table F.1. Type 1 and Type 2 experiments, Fertilizer Use Recommendations Project, Kenya.

	Type 1 experiment		Type 2 experiment			
Treatment number	P (kg ha^{-1})	N (kg ha^{-1})	N and P (50 kg ha^{-1} each)	K (50 kg ha^{-1})	Lime (500 kg ha^{-1})	FYM (5 t ha^{-1})
1	0	0	0	0	0	0
2	0	25	0	0	0	1
3	0	50	0	0	1	0
4	0	75	0	0	1	1
5	25	0	0	1	0	0
6	25	25	0	1	0	1
7	25	50	0	1	1	0
8	25	75	0	1	1	1
9	50	0	1	0	0	0
10	50	25	1	0	0	1
11	50	50	1	0	1	0
12	50	75	1	0	1	1
13	75	0	1	1	0	0
14	75	25	1	1	0	1
15	75	50	1	1	1	0
16	75	75	1	1	1	1

Note: N = nitrogen; P = phosphorus; K = potassium; FYM = farmyard manure. Zero indicates no application and 1 indicates that the treatment is applied (KARI 1990).

Reference

KARI (Kenya Agricultural Research Institute) (1990) *The Fertilizer Use and Recommendations Project (FURP): Main Report.* National Agricultural Research Laboratories (NARL), Nairobi, Kenya.

Appendix G
The Transcendental Function

In log form:

$$\ln Y = \ln A + \Sigma_i \ \ln X_i + g(x),$$

where the general form for $g(x)$ is:

$$g(x) = \Sigma_i \ \beta_i \ X_i,$$

extended in our case to be:

$$g(x) = \Sigma_i \beta_i X_i + \Sigma_i \Sigma_j \beta_{ij} X_i X_j + \Sigma_l \ dl \ D_1 \ + \Sigma_i \Sigma_1 \ d_i l \ D_1 X_i,$$

where Y is yield and X_i measures the level of continuous variable i. The set of regressors contains continuous as well as discrete factors (D). The specification of $g(x)$ used in this study allows for interaction effects as measured by β_{ij} and $d_i l$, where: β_{ij} measures the interaction effect between two continuous variables i and j; D_1 is a dummy for the effect of categorical variable l; dl measures the deviation of the mean effect of category l from the overall mean effect (common intercept); and $d_i l$ measures the interaction between continuous (X_i) and categorical variables (D_1). In other words, $d_i l$ measures the deviation of the marginal effect of variable i for category l from the overall effect (common slope) of variable i.

Index